IF YOU THINK YOU KNOW THE WHOLE STO[...]
READ COLM A. KELLEHER'S NATIO[...]

BRAIN TRUST

The Hidden Connection Between Mad Cow and Misdiagnosed Alzheimer's Disease

—AND DISCOVER THE SHOCKING SCIENCE BEHIND A GROWING EPIDEMIC, THE TRUTH BEHIND A DEVASTATING COVER-UP, AND THE IMPLICATIONS FOR YOUR HEALTH!

HUNT FOR THE SKINWALKER

SCIENCE CONFRONTS THE

UNEXPLAINED AT A

REMOTE RANCH IN UTAH

Colm A. Kelleher, PhD, and George Knapp

PARAVIEW POCKET BOOKS

New York   London   Toronto   Sydney

PARAVIEW
191 Seventh Avenue, New York, NY 10011

POCKET BOOKS, a division of Simon & Schuster, Inc.
1230 Avenue of the Americas, New York, NY 10020

Library of Congress Cataloging-in-Publication Data

Kelleher, Colm A.
    Hunt for the skinwalker : science confronts the unexplained at a
    remote ranch in Utah / Colm A. Kelleher and George Knapp.—1st
    Paraview Pocket Books trade pbk. ed.
        p. cm.
    Includes bibliographical references (p.) and index.
    1. Parasychology—Utah—Case studies. I. Knapp, George. II. Title.

BF1028.5.U6K45 2005
001.94'09792—dc22   2005053457

ISBN-13: 978-1-4165-0521-1
ISBN-10:   1-4165-0521-0

First Paraview Pocket Books trade paperback edition December 2005

20 19 18 17 16 15

For information regarding special discounts for bulk purchases,
please contact Simon & Schuster Special Sales at 1-800-456-6798
or business@simonandschuster.com.

To Robert Bigelow, for his brilliant and farsighted vision.

To the Gorman family, for their extraordinary courage.

Contents

PART III: AFTERMATH AND HYPOTHESES

25.   Hypotheses   206
26.   The Military   220
27.   The Native American Connection   229
28.   Other Dimensions   237
29.   Outer Worlds   242
30.   Inner Worlds   249
31.   Revolutionary Science   257

      Epilogue   271
      References   283
      Acknowledgments   293
      Index   295

Preface

This book is an account of a remarkable series of unexplained events that took place on a ranch in northeastern Utah and the unprecedented scientific study that followed. For eight years, a team of highly trained scientists and others came face-to-face with a terrifying reality that, on superficial examination, appeared to break the laws of science but that, in fact, was consistent with modern-day physics. At the ranch, scientists found a world where a great deal of activity was hidden from visible sight but, as the researchers soon discovered, was detectable with state-of-the-art instrumentation. The family that lived there came to believe that the ranch was occupied by some kind of intelligence that appeared to control—as if on a whim—human perception, human thought, and human physical reality.

The ranch in question lies well off the beaten path in a remote, rural corner of Utah, but is only about 150 miles from metropolitan, sophisticated Salt Lake City. The location is in the midst of a devout Mormon community and is contiguous with a Ute Native American reservation. Both communities have experienced unbelievable but well-documented phenomena in their midst for more than fifty years. In the case of the Ute tribe, the experiences are documented in the tribe's oral tradition stretching back over fifteen generations.

The account you are about to read is true. I know because I directly participated in and witnessed several of these events myself. All of these incidents really happened. I, along with a small team of highly trained scientists and investigators, interviewed hundreds of eyewitnesses to these strange occurrences, including law enforcement officers, physicists, biologists, anthropologists, veterinarians, educators, and everyday citizens.

In addition to eyewitness testimony, we obtained an intriguing body of physical evidence to support many of the accounts described in the book. We compiled photos and videos and accumulated reports of demonstrable physical effects on people, animals, equipment, everyday objects, and the environment. Although observers might relegate the subject matter to the category of the paranormal, the research team adhered to the strictest scientific protocols throughout the project.

Some of the names in the book have been changed out of concern for the emotional well-being of the family that owned the ranch at the time and for the sake of others who were involved. As readers will discover, this family endured a painful and disturbing series of events that left deep psychological scars. The family has since moved from the ranch and is trying to put these events behind them. We also omitted the names of a physicist and a veterinarian, out of concern that the strange subject matter they pursued at the ranch might interfere with their ability to obtain future employment. The scientific establishment does not look kindly upon professionals who stray too far from what are deemed legitimate areas of study.

We also do not provide the exact location of the ranch itself. We are concerned that specific information about how to find it would encourage intrusions and trespassing by curiosity seekers and paranormal enthusiasts, which has already occurred to some extent. The ranch itself is still private property and its caretakers do not welcome incursions by strangers. Neighbors in this rural area also do not appreciate knocks on their doors from out-of-towners seeking paranormal thrills and other strange experiences. That said, the reader will learn about the region, the towns near the ranch, and the geography of the ranch itself, including specific information about where various events occurred.

An advisory board of esteemed scientific professionals oversaw this study of the ranch. This board was probably the most highly qualified team of mainstream scientists ever to engage in such a sustained study of anomalies. Board members insisted that established scientific principles and procedures be followed for the duration of the study. The problem is that we were forced to engage *someone* or *something* that refused to play by the rules of science. As a consequence, I realized that, despite my training in the minutiae of experimental protocols in immunology, biochemistry, and cell biology, we had to creatively modify the tried-and-true methods of establishing scientific experimental controls and of working under controlled laboratory conditions. We were obliged to conduct science in a weird shadowy netherworld where textbook science was but a quaint memory.

The account you are about to read is purely my own; it is not meant to represent the views of my employer at the time, or of the other members of the research team. I believe that these strange, sometimes frightening, often bewildering events represent far more than a potpourri of unrelated and unfathomable weirdness. In the end, I suspect that this intense concentration of "paranormal" activity could point us all toward a new understanding of physical reality, something that is already being debated at the highest levels of modern science.

The world, it appears, is much bigger, much stranger, and far more complicated than most of us can imagine.

Colm Kelleher

A scientist or mainstream journalist who decides to give serious consideration to unidentified flying objects or other so-called paranormal topics does so at considerable risk to his or her professional standing. I learned this the hard way.

In my twenty-five years as an investigative reporter, TV anchorman, and newspaper columnist in one of the world's most dynamic cities, I have been fortunate to cover stories large and small. I've tangled with Mafia figures, professional hitmen, casino moguls, crooked politicians, drug dealers, gunrunners, car bombers, arsonists for hire, porn kings, outlaw motorcycle gangs, dirty cops, illegal polluters, animal abusers, bagmen, con men, scam artists, pimps, perverts, and scumbags of every stripe. In my own community, I'm generally regarded as a serious journalist. I mention this not as a boast but as a reference point. Even though I've written reams of stories about every imaginable topic, it seems likely that I will forever be known in my hometown as the UFO reporter. In the eyes of many of my journalism colleagues, this means I'm a borderline nutcase.

I officially went crazy in 1989. That's when I produced a multipart television news series about UFOs and a mysterious military base the world now knows as Area 51. The viewing public ate it up. But my fellow journalists were not pleased. For some reason, this serious coverage of a "fringe" topic drove them up the wall. I was pilloried and ridiculed from every quarter.

Rival TV broadcasters snickered and lobbed snide potshots during their "happy talk" segues. Radio DJs made crank calls and recorded song parodies. (The Beatles' "Fool on the Hill" was transformed into "Boob on the Tube.") Newspapers were relentless in their criticism. Columnists teed off with predictable wisecracks about Bigfoot, Elvis, and the need for ET to phone home. A media critic wrote that he was rushing home each night to see my reports, not because of his interest in UFOs but so he could be there for the inevitable moment when I finally went "bull-goose loony on the air." One newspaper writer reacted to my reports about alleged alien abductions by anointing me as "a high priest in the Church of Cosmic Proctology." The same editorial staff later opined that my interest in UFOs besmirched the reputations of other investigative journalists, so the staffers awarded me the dubious distinction of being the city's "Biggest Blowhard," which in a town like Las Vegas is quite an accomplishment.

I was the subject of three editorial cartoons in Nevada's largest newspaper, cartoons that were, I have to admit, pretty damned funny. One portrayed a portly likeness of me chasing a flying saucer while brandishing a butterfly net. Another featured alien assassins who had arrived on Earth aboard an interstellar barbecue grill for the purpose of rubbing me out. The subtitle of the cartoon was "The Marshmallow Head Chronicles." I didn't have to guess whose head was considered to be soft and spongy.

My general reaction at the time was that criticism of this sort comes with the territory. Whining about it would be pretty pathetic. A person can't be in the public eye and expect to get a free pass, especially from competitors. But the vehemence with which so many of my journalism colleagues attacked me was a surprise. After all, I was the same guy who had produced so many other stories. How did I morph into the crazy UFO reporter?

The truly puzzling part was that none of these critics had even a basic familiarity with the large body of UFO evidence, studies, and documents. Their impression of the topic was—and is—based largely on a generic, ill-defined belief that people who are interested in flying saucers are delusional. (Some are, of course, but the same could be said about many journalists.) This impression has been reinforced over the years by the ridiculous tabloid accounts of space aliens who regularly visit the White House or the goofball claims of people who think they were born in another galaxy. In general, mainstream journalists have been reluctant to look beyond the exaggerated fictions of publicity seekers and profiteers in order to find out if there really is something going on. To do so would be to put their professional credibility on the line. I hope that those mainstream journalists who read this book will suspend their disbelief, at least temporarily, because the subject matter deserves serious inquiry, in my opinion.

One of the best things to stem from my unanticipated status as the UFO guy was the opportunity to meet the remarkable Bob Bigelow and, later, members of the board of the National Institute for Discovery Science and its staff scientists, especially Colm Kelleher. Whatever sniping and peer pressure I encountered from journalism colleagues, it was negligible compared to the professional scorn and career consequences that loomed for Kelleher and the rest of the NIDS team. Real scientists simply do not participate in crackpot research projects. A reporter who uncorks a bizarre story or two can eventually be forgiven. A scientist who chases after UFOs or mutilated cattle risks everything. It shouldn't be like this, but it is. To me, it makes the NIDS study of the Utah ranch even more astounding.

Although I was a mere observer to the events that unfolded, the courageous investigation by Bigelow, Kelleher, and the other NIDS team members has shaken me to the core. I will never look at the world in the same way. As readers are about to learn, reality isn't what it used to be.

George Knapp

PART I

THE HOTSPOT

CHAPTER 1

"Wolf"

What is that?" Tom Gorman wondered as he looked across the field at the distant animal loping in his direction. He paused briefly and put down the heavy box he had lifted off the truck. Tom had the perfect eyesight of a trained marksman, and he knew from a half a mile away that this animal was big. The approaching shape was much too big for a coyote. His wife, Ellen, joined him, an unspoken question in her eyes. Tom briefly nodded his head in the direction of the animal and she too began to look puzzled. The thing was about four hundred yards away, and the closer it got the bigger it looked. "Wolf?" murmured Ellen. Ed Gorman, Tom's father, joined them.

The beast was gray, and even from three hundred yards they could see that its pelt was wet from running through the wet grass. The animal loped gracefully in a series of S turns and stopped about fifty yards away from the family. This was very bizarre behavior for a wolf. But this wolf was almost three times as big as any Tom had ever seen. It gazed peacefully at the family.

Ellen shifted uneasily and glanced around to see where her two children were. Both were standing in silence on the top of the flatbed truck, looking right at the wolf. "Maybe its somebody's pet," ventured Ed.

The animal began walking casually toward the family, unconcerned but obviously determined to make some kind of contact. It appeared completely tame. Tom glanced at the corral seventy feet to his right where he had just unloaded several of his prize Angus calves. They were the first of his herd to be on the property, and briefly he wondered about the wisdom of bringing them onto the land. One calf, more curious than the rest, stood with its head through the bars of the corral, looking directly at the wolf that was now only a hundred feet away. The other animals were at the back of the corral, and they shifted nervously at the strong scent in the air.

From ten feet away, the smell of rain on dog pelt filled the air as the animal trotted peacefully up to Ed Gorman. Ed, like his son, stood over six feet tall, and the wolf reached almost to his chest. Massive muscles rippled beneath its shiny gray-silver coat. The eyes were a shocking shade of light blue that penetrated the soul. Ed reached down and petted the huge beast as it stood looking at the family. Tom felt a tightening in his gut. Something was not quite right. Even somebody's pet wolf would not be this completely tame. A two-hundred-pound wolf exuding a Zen-like calm? Something did not compute.

The animal walked nonchalantly around in front of the family, and Ellen and Ed began to relax. Ellen turned around and yelled to the kids to come over. Tad and Kate Gorman jumped from the flatbed and ran over. The family began talking all at once. Tad suggested they try to keep the wolf as a pet.

Too late, they saw the swift, graceful bound that took the wolf to the bars of the corral. With unbelievable speed, the young calf's head was trapped in the animal's powerful jaws. The movement had been lightning fast, and the family stood paralyzed with fear. The three-hundred-pound calf bleated pitifully as the wolf tried to drag it through the bars of the corral. Tom sprang into action, ran across, and landed two powerful kicks into the ribs of the wolf. Ed followed and grabbed a stout baseball bat he had just unloaded. With all of his considerable strength, Ed beat on the wolf's back as it braced against the bars of the corral trying to drag the hapless calf through it. The bleats were getting more urgent as the viselike lock on the calf's snout tightened.

"Get my Magnum," Tom barked as he aimed more kicks at the wolf's ribs. Even as the sickening thud of Tom's heavy boots rained into the animal's abdomen, the beast seemed unconcerned. Tad ran to the flatbed, retrieved a powerful handgun, and quickly delivered it to his father. Gorman took aim and squeezed the trigger. The shot rang across the field and slammed into the wolf's ribs. The slug from the .357 had no effect whatsoever on the attacking animal. It didn't yelp, didn't pause, and didn't bleed. Quickly, Tom pumped two more shots into the wolf's upper abdomen. On the third shot, the wolf slowly and reluctantly released the bleating calf. The calf scampered quickly to the back of the corral and, still bleating, lay down. It was bleeding heavily from the head.

The huge beast stood about ten feet away from Tom but displayed no signs of discomfort. Tom couldn't believe it. Three shots from a Magnum should have killed the animal or at least very badly injured it. Not a sound came from the wolf as it gazed unconcernedly at Gorman. The chilling, hypnotic blue eyes looked straight at him. Gorman raised the Magnum again and, aiming carefully, shot the animal near the heart. It backed off maybe thirty feet, still facing the family and still showing no signs of distress.

A chill crept over Tom. The family drew closer together. They were all more than familiar with the power of the Colt Magnum. They had seen firsthand the devastation it causes, yet this huge wolf was not even making a sound after being shot four times at point-blank range. There were no signs of blood on the beast. It seemed peaceful but glanced back at the calf in the corral as if pondering the wisdom of another attack.

"Get the thirty aught six," Tom said through clenched teeth, never taking his eyes off the huge beast. Tad ran to the homestead and returned in seconds bearing the heavy firearm. Tom had killed dozens of elk from over two hundred yards with this weapon. As he took aim at the wolf a mere forty feet away, he momentarily felt pity for the beast. The thunderous shot rang out. The sound of the bullet hitting flesh and bone near the shoulder was unmistakable. The wolf recoiled but stood calmly looking at Tom. His mouth went dry. He felt a cold sweat running down his back. Ellen began to cry. Ed began to curse softly under his breath, shaking his head in disbelief. The wolf should be a silent, bleeding pile of dead flesh. Instead, it had recoiled, backed off maybe ten feet, but still seemed perfectly healthy.

Tom took a deep breath and raised the weapon again, aiming for the huge chest cavity. The bullet ripped through the animal, and a sizable chunk of flesh detached from the exit wound and lay on the grass. Still the wolf made no sound. Then, with a last unhurried look at the stunned family, the wolf turned slowly and began to trot away across the grass. Tears of fear streamed down Ellen's face as she hugged her twelve-year-old daughter.

Tom's face was white and there was strain in his voice as he turned toward his family. "Let's keep calm," he muttered hoarsely but didn't sound very convincing. "I'm going after it." The animal was now almost a hundred yards away, trotting west across the field in the direction of a dense group of cottonwoods. Beyond the cottonwoods lay a roaring creek. Tad grabbed the Magnum and Tom hefted the thirty aught, and the family watched as they sprinted off in the same direction as the wolf. The animal was only trotting but was covering ground quickly.

Anger and fear pulsed through Tom as he pushed himself to

run quicker. He was already out of breath, but they were gaining on the wolf. They could see the animal disappearing into the belt of cottonwoods and then reappear in the open ground beyond. It stopped, momentarily shaking itself free of the moisture from the grass before heading for the creek. Tad ran silently, feeling how upset his father was but concentrating on keeping the wolf in his sight. The wolf seemed to be accelerating. It was now almost three hundred yards ahead of them and still loping easily as it reached a denser patch of Russian olive trees that bordered the creek.

As they ran, Tom noticed that the tracks of the animal were easily visible in the wet ground. Gorman was an experienced tracker and he was confident that they could track the animal even in the thick Russian olives. His sharp eyes spotted the silvery-gray blur as the animal disappeared into the tree line. Minutes later, Tom and Tad ran into the line of trees following the giant animal's tracks. In some places it had left inch-deep impressions in the soft ground. There was no evidence of blood on or between the huge footprints.

Tom couldn't shake the fear he felt as he brushed through the tightly woven undergrowth. His pace had slowed because the large trees were interwoven with thorny brambles and weeds. The tracks were still visible. As they approached the creek, they could hear the water gurgling as it cascaded merrily over the rocks.

They broke cover near the bank of the creek and Tom held up his hand. Tad stopped and the two listened carefully. They heard no sound of an animal crashing through the undergrowth. The huge paw marks periodically meandered in and out of the surrounding vegetation but consistently shadowed the direction of the creek. Tom guessed they had run about a mile.

Several minutes later the two broke through into the open about forty yards from the river. They breathed a sigh of relief. It was hard going, stumbling through the trees, making sure the head-high thorns and bushes didn't take a toll on their skin and their faces. Suddenly, Tom stopped breathing. He grabbed Tad's arm and pointed. The wolf tracks were directly in front of them, as plain as day, as they headed toward the creek. About twenty-five yards from the river, the prints entered a muddy patch, and it appeared as if the two-hundred-pound animal had sunk almost two inches into the mud. The deep paw prints continued for another five yards and then stopped. The tracks simply vanished. So did the wolf. Gone. There was no possibility that the animal had leaped the intervening sixty feet to land in the river. The tracks just stopped abruptly.

The Gormans walked slowly and carefully, looking at the perfectly formed tracks in the thick mud and trying to see any change that might explain the sudden disappearance. Around where the tracks halted, the ground appeared about as soft as the mud patch. It was as if the animal had vanished into thin air. Tom looked at his son, and he could see that the teenager was white faced and trembling, close to tears. Tom felt stunned. He couldn't reassure his son, because he just didn't have an explanation. "We'd best be getting back," he said hoarsely. "It's near sundown." Tad nodded dumbly, fighting to keep his father from seeing how scared he was.

They were silent as they trudged the miles back to the homestead. Thoughts raced through Tom's head. The family had just moved from New Mexico to get away from the busybodies and the closed community that kept prying into their lives. They had looked in Utah because property prices were right. In this out-of-the-way place, tucked away in northeast Utah, they had found their dream property—a 480-acre homestead that had been empty for almost seven years. The elderly previous owners had virtually abandoned it. The owners were a prosperous family who now resided in Salt Lake City, and they visited their property a couple of times a year to make sure the fence lines were intact. They were willing to unload the property to the Gormans at a very fair price. The family knew that about a year of hard work would be required to fix it up. High ridges bound the acreage to the north, the flowing creek to the south, and extensive fencing to the west. The homestead was hidden about a mile from the nearest road, down a dirt track that was almost concealed. In short, it was a perfect refuge for a family that yearned for privacy and a home where they could relax and put down roots. The Gormans were happy to trade a small-town life in New Mexico for a new start in a Mormon community in rural Utah. Like most of their new neighbors, the Gormans were members of the LDS church, although they could not be considered devout.

As they trudged through the deep undergrowth, Tom couldn't shake the feeling that something was horribly wrong. Had they made a mistake in buying this place? Thoughts tumbled through his mind, causing his gut to tighten even more. He knew something had happened today that everyone knew was physically impossible. And it had occurred in daylight and in full view.

Quickly, Tom came to a decision as he stood facing his family. He was not going to second-guess their decision to move from New Mexico. "Look, son," he said to Tad. "I can't explain what happened and I am not even going to try. Let's just forget that this ever happened and have a meal in town."

Tad just grinned weakly, relieved at least that his father was taking charge.

CHAPTER 2

Legacy

The first time they saw the ranch, its beauty took their breath away. The Gorman family drove the half-mile track into the yard near their small homestead and marveled at the sheer pastoral magnificence: 480 acres of cottonwood trees, Russian olives, and very lush pasture bordered by a creek, and an irrigation canal bubbling near the northern limit of the property. The family was entranced when they first explored this idyllic spot. Later, they would learn of its drawbacks.

The ranch is bordered on the north by a two-hundred-foot ridge made of red rocks and the mud derived from centuries of weathering. When it rains, the mud below the ridge becomes a thick, slippery mess. And right next to the canal, a muddy track runs the entire east-west length of the property. This too becomes a nightmare to drive on after even a slight rainfall. The Gormans were reminded many times never to drive on that track when it began to rain. They had to haul all of their vehicles out of the canal several times before they finally learned that lesson.

The Gormans couldn't believe their luck as they first walked the length of the property. They had purchased it for a very fair price. It was the fall of 1994, and many of the leaves were still on the trees. Just beside their homestead, a large pasture that contained a lot of rocks and trees stretched for almost half a mile west. It needed a lot of work, but the spectacular view shone through the disused property that was littered with garbage. Tom realized that the coming months would be tough and much work would be needed before he could bring his herd of prize, registered cattle to graze on the ranch.

When the Gormans first entered the small ranch house that was to be their home, they felt a chill. Every door had several large, heavy-duty dead bolts on both the inside and outside. All of the windows were bolted, and at each end of the farmhouse, large metal chains attached to huge steel rings were embedded securely into the wall. The previous owners had apparently chained very large guard dogs on both ends of the house. And they had barred the windows and put dead bolts on both sides of each door. What on earth were they afraid of? Tom wondered.

The previous owners had bought the property in the 1950s but now seemed glad to unload it. They had inserted some very strange clauses into the real estate contract. No digging on the land without prior warning to the previous owners. No digging? What did that mean? The Gormans chose to overlook this seemingly insignificant idiosyncrasy, regarding it as a meaningless clause crafted by elderly eccentrics. They put it out of their minds, in the same way they had glossed over the profusion of dead bolts in the house, which could have easily been interpreted as evidence of extreme paranoia on the part of the former residents.

The Gorman family was more intent on savoring the beauty of the environment, the out-of-the-way location of the ranch, and the certainty that they could raise their children in rural surroundings where the value of hard work and family life would supplant the small-town sniping and gossip that they so loathed in New Mexico. Tad and Kate had always been straight-A students. They took after their parents, who were extremely intelligent, diligent, and hardworking. Tom Gorman was a very accomplished rancher who combined common sense with a razor-sharp intellect. He also had a preternatural sixth sense and operated by intuition. That intuition, and his intellect and common sense, combined to create a very capable rancher, a man who could excel at most anything he did. His wife united her high intelligence with a natural business sense.

The Gorman family had learned the rudiments of artificial insemination by watching Tom's dad do it and by training. Even before moving his family to northeast Utah, Gorman had already established a reputation in a couple of states as an expert in raising the top-quality Simmental and Black Angus show cattle that fetched especially high prices at cattle auctions. While their neighbors routinely lost 5 percent of their animals every year to predators, incompetent husbandry, and other mistakes, the Gormans saw it as a personal affront if they lost more than 1 percent of their animals per year.

The ranch the Gormans bought was located in the middle of the Uinta Basin (the *h* in Uintah is dropped when describing natural features), halfway between Roosevelt and Vernal in the badlands of Utah. If the Uinta Basin has any claim to fame today, it's as a heavyweight contender for UFO capital of the world. Since the 1950s, thousands of UFO sightings have been reported in the area, and it easily ranks among the most active UFO areas anywhere. By some estimates, more than half of the residents of the basin have seen anomalous objects in the sky. Admittedly, 90 to 95 percent of so-called UFO reports are misidentifications of known phenomena. But even by that rigid standard, a very large number of sightings in the Uinta Basin can be categorized as unexplained.

In fact, in 1974, Frank Salisbury, then a plant physiologist and a professor of plant science at the University of Utah, wrote a well-received book on the history of the UFO phenomena in the Uinta Basin titled *The Utah UFO Display*. Salisbury compiled a very convincing case that local residents were witnessing something very weird, indeed, sensational. Unlike most UFO authors, Salisbury refrained from wild speculations about little green men from Zeta Reticuli. He stuck to the facts because the facts were sensational enough. But Utah's UFO wave did not end with Salisbury's study. It continues, unabated, to this day.

"The first UFO sighting in my records happened back in 1951," says retired teacher Junior Hicks, a small, energetic, wiry man now in his seventies who is widely acknowledged as the region's unofficial UFO historian. "It was cigar shaped, sitting on the ground during daylight, and was seen by thirty students and their teacher from about fifty feet away."

Hicks, as a science teacher, took it upon himself to follow up on the case and separately interviewed all of the kids who had seen the object. He quickly concluded that the kids had not made it up. The case so intrigued him that he began to actively pursue other reports of UFOs in the basin. Typically, he would contact witnesses at their homes, arrange to meet them in a quiet and comfortable setting, then interview them, face-to-face, for several hours. Hicks never divulged the identity of the witnesses without their permission. As word of his trustworthiness spread, Hicks began to receive more and more calls about the mysterious objects that seemed to hold a fascination for the Uinta Basin.

Hicks would eventually catalog more than four hundred impressive cases, and this was after he had eliminated the thousands of reports of "lights in the sky." Hicks's database was heavily skewed toward close encounters simply because he was too busy to investigate anything but the more spectacular cases. Hicks's case files helped form the core of Salisbury's book.

This strange legacy of the Uinta Basin goes back centuries,

Hicks told us in an interview in 2003. "Father Escalante may have seen a UFO when he was here in 1776," he said. "The records from his trip show that while encamped at El Rey, a strange fireball came across the sky above his camp. The UFOs seen here since I've been collecting stories range in size from twenty to thirty feet across, all the way to the size of a football field. Some are round, some oval, some cigar shaped, some triangular. The largest one, a triangle, was seen back in the sixties. We had one resident, an Indian, who took a shot at a UFO with his rifle and heard a ricochet ping as the bullet bounced off the metal ship. The people who see them include lawyers, bankers, ranchers, people I've known my whole life."

Because he taught school for thirty-six years, Hicks has a personal relationship with most of the people who live in the area and has been able to talk openly with them about their experiences, whereas some outside investigators who've asked questions have been stymied by mistrust among the local residents, many of whom are understandably wary of being ridiculed by big-city strangers. Hicks says there was a time in the 1960s and 1970s when the Utah Highway Patrol was getting so many UFO calls that the troopers simply stopped filling out reports on the incidents. The UFO stories have predictably attracted a steady stream of journalists, TV crews, and UFO faithful over the years, but they generally stay for only a few days, he says, and the reports they produce rarely do more than skim the surface of what is really unfolding in the basin.

Hicks had his own sighting back in the mid-1970s. He watched an orange ball fly over the town of Roosevelt at a high rate of speed, then make an abrupt right-angle turn. The ball hovered in the air over the town before zipping out of sight at an incredible speed. In at least six of the cases he investigated, the witnesses say they saw not only spaceships but also the occupants of the craft. A rancher whose father had been a Native American shaman told Hicks that a silver saucer landed on his land and that five short human-looking beings could be seen walking around inside the craft. The saucer had a row of windows, the witness said, and the beings inside appeared to be wearing white overalls.

So why would UFOs have such an ongoing interest in the basin? Hicks thinks that it might be related somehow to strong religious beliefs among the local population. The majority of Uintah County residents are devout Mormons, and the Utes who live in the region have their own strong religious beliefs. Hicks also cites the unique geology of the basin as a possible factor in the preponderance of UFO sightings. Gilsonite deposits, for example, are found in few other places in the world, although it's unclear what interest the mysterious visitors might have in this or any other mineral. Hicks notes that the basin is biologically rich and diverse. At least one UFO witness he interviewed claims to have seen alien beings as they collected samples of native plant life.

In the end, though, he suspects the continued presence of the UFOs might be some sort of psychological exercise designed to see how humans might react to these bizarre light intrusions. "They seem to put on displays for people," Hicks says, "as if it's a psychological study. I think we are being visited by beings from another world or some other place and it's for research and exploration." Hicks has investigated two incidents in which local residents claimed to have been abducted by aliens.

But along with the aerial displays by UFOs have come more frightening intrusions. Beginning in the 1960s and continuing into the 1970s, local ranchers reported the bizarre mutilations of their cattle. Similar mutilations have been reported in other parts of the world. The motives and methods of the mutilators, none of whom have ever been caught in the act, remain a complete mystery. Hicks says he has personal knowledge of twelve to fifteen mutilation cases.

As if checking off a laundry list of every modern paranormal mystery, Hicks further alleges that local residents have frequently reported sightings of creatures that resemble the legendary Sasquatch, better known as Bigfoot. Some of the apelike creatures are Sasquatch, the Utes say, while others might be so-called skinwalkers, beings of pure evil that can assume the shape of any animal.

Despite the many reports of UFOs, animal mutilations, and Bigfoot sightings that seem to permeate every corner of the basin, the greatest concentration of high strangeness has always taken place at what became the Gormans' 480-acre ranch. Junior Hicks says he has worked on the ranch a few times over the years. He helped to repair pumps and performed other small jobs, and during those visits, he and others have seen things that are not easily explained. He's seen compasses spin wildly out of control, as if disturbed by unknown magnetic forces.

"It all seems to be concentrated on the ranch," Hicks says. "The Utes don't mess with it. They have stories about the place that go back fifteen generations. They say the ranch is 'in the path of the skinwalker.' "

CHAPTER 3

The Basin

The Uinta Basin has always been more than a little strange.
When Mormon leader Brigham Young sent a small expedition to the basin in the 1860s to see if the region was suitable for settlement, the reviews weren't favorable. The scouting party reported back to Young that the basin was "a vast contiguity of waste . . . valueless except for nomadic purposes, hunting grounds for Indians and to hold the world together."

The Uinta Basin is a challenging environment, but a wasteland it isn't. This stunningly beautiful geographic area of northeastern Utah has been inhabited for more than twelve thousand years by Native American tribes. The first white men to visit the region arrived with Spanish expeditions led by Fathers Dominguez and Escalante in the 1770s. Hunters, trappers, and traders followed. In 1861, President Abraham Lincoln established the Uintah Indian Reservation that encompassed most of the basin. The action was taken because of the numerous armed conflicts between Utes and Mormon settlers in the Provo Valley.

Lincoln's order meant that the Utes had to leave the greenery of the Provo Valley for the harsher environment of the Uinta Basin, 150 miles to the east. The tribe was promised that the reservation lands would belong to the Utes for all time. But within a few years, white settlers began to covet those lands for their own economic pursuits, so the boundaries of the Ute territory were slowly and inexorably carved into ever-smaller pieces.

When Mormon settlers returned to the Uinta region to stay in the 1880s, the Utes' hold on their lands became even less tenable. In 1885, miners prospecting on tribal lands discovered rich deposits of a black hydrocarbon that would prove essential in the manufacture of paints, varnish, lacquer, and insulating materials. The rare mineral was named gilsonite, after Samuel Henry Gilson, an early proponent of gilsonite's economic potential. When word spread about the value of gilsonite, aggressive mining interests staked claims and set up operations on reservation lands, even though such actions were clearly illegal. An honest Indian agent named T. M. Byrnes temporarily forced the brazen miners to shut down. The mining companies petitioned Congress to declare that more than seven thousand acres of gilsonite-rich Indian land should be reclassified as "public domain." Since the property rights of Indian tribes weren't a high priority, Congress approved the bill. The Utes were to be compensated with payments of twenty dollars per acre. Those tribe members who didn't want to sell were plied with whiskey or otherwise tricked, and by 1888 the mining interests obtained control of all of the land they originally sought.

In 1881, another reservation was established adjacent to the first in order to accommodate bands of Utes that were forced by the government to get out of Colorado. This concentration of so many potentially hostile Indians prompted the military to authorize the establishment of a new military outpost, one that would be responsible for guarding the Indian frontier in eastern Utah, western Colorado, and southwestern Wyoming. In August of

1886, Major Frederick Benteen led a contingent of the Ninth U.S. Cavalry to the spot in northeastern Utah where the Duchesne and Uinta rivers met. Ten years earlier, Benteen had served with General George Armstrong Custer's ill-fated Seventh Cavalry. Benteen survived the massacre at Little Bighorn and eventually was given the assignment of establishing a military outpost at a godforsaken juncture of two obscure rivers in a territory that, for centuries, had been the exclusive domain of the Utes, a nation of proud, fearsome, and unpredictable warriors.

When Benteen rode into what would eventually become Fort Duchesne, seventy-five battle-tested cavalrymen accompanied him. Every one of the seventy-five troopers was black, the legendary Buffalo Soldiers. The 150 or so white infantrymen who had marched into the area a few days earlier cheered the arrival of reinforcements. By some accounts, the Utes who witnessed the entrance of the new arrivals were less enthusiastic. The reputation of the Buffalo Soldiers preceded them.

The legacy of the Buffalo Soldiers is well documented in history books. In the bloody campaigns of the Indian Wars in the late 1800s, approximately 20 percent of U.S. Cavalry troopers were African Americans. Native tribes dubbed them Buffalo Soldiers, in part because of a perceived resemblance between these dark-skinned, curly-haired warriors and the revered buffalo, and in part because of their prowess in battle and in the saddle. At least eighteen Congressional Medals of Honor were awarded to Buffalo Soldiers as a result of their actions in at least 177 armed engagements during the Indian Wars in the West.

What isn't widely known about the Buffalo Soldiers stationed at Fort Duchesne is that many, if not most, of them were Freemasons. This may not seem consistent with our current perception of Masons or Freemasons as mostly white, mostly upper-class captains of industry and politics, but it happens to be true.

Freemasonry is a secret society that traces its roots back to ancient Egypt. The society first appeared in Europe in the 1300s after a predecessor, the Knights Templars, fell into official disfavor and was generally outlawed. The grandfather of today's Freemasons made its public debut in London in the early 1700s and declared itself a fraternity whose mission was to promote charity and the improvement of society. However, its reliance on secret handshakes, secret ceremonies, its own sign language, and overt mystical symbolism has prompted centuries of suspicion by non-Masons.

Critics of the society—and there are many, practically a society unto themselves—note that the early Freemason lodges in England were dominated by alchemists, astrologers, and students of the occult. The society's history of absolute secrecy has given birth to generations of lurid speculation, including allegations that the Freemasons are servants of Lucifer. The Masons have done little in the centuries since to rid themselves of their perceived reputation as a festering cauldron of dark intrigues and the mystical arts.

In America, many of the key figures in the formation of the republic were Masons, including Benjamin Franklin. (George Washington attended two or three meetings but wasn't a cheerleader for the cause, whatever that cause might be.) American currency is plastered with Masonic symbols and slogans, largely because of the efforts of a thirty-third-degree Freemason named Franklin D. Roosevelt. Future president George H. W. Bush gained his entry to the club when, as a student at Yale, he was initiated into an alleged Freemason farm team known as the Skull and Bones Society, which accepts a maximum of fifteen new members each year, all of them male and all of them rich.

Conspiracy theorists suspect that the Freemasons are adept in the mystical arts, that they have mastered certain supernatural abilities, and that they consider their upper-echelon members to be gods, beings who have achieved spiritual perfection. On a more mundane level, critics also allege that the Freemasons, or their alleged co-conspirators the Illuminati, the Trilateralists, and the Bilderbergers, are intent upon imposing a new world order, a one-world government, a system under which national interests are subservient to the greater planetary good, that is, "good" as determined by the Freemasons. This is one hell of an ambitious conspiracy theory.

It seems so incongruous on the surface to presume any possible connection between black soldiers stationed at a remote outpost in Utah in the late 1800s and a secret, mystical society still bent on world domination in the early twenty-first century. Nonetheless, the Buffalo Soldiers of Fort Duchesne were full-fledged, ritual-practicing, secret-handshaking members of the world's best-known, most influential, and most mysterious male fraternity.

The soldiers of Fort Duchesne became Freemasons because of an emancipated slave named Prince Hall who emigrated from England to America and established African Lodge No. 1 in Boston on July 3, 1776, certainly a precipitous moment in American history. Although there are disputes about the legitimacy of this account, "Prince Hall Lodges" quickly multiplied in the new nation. One of those lodges was established in Texas in the mid-1800s, which is where it intersected with attachments of the Ninth and Tenth Cavalry, the Buffalo Soldiers. A few decades later, the Masonic seeds that had been planted by that frontier lodge in Texas found their way to a remote corner of Utah.

The Utes who live in Fort Duchesne today are very familiar with the stories about the Buffalo Soldiers and their interest in Freemasonry. A patch of ground that once was designated as the graveyard for the Buffalo Soldiers has since been covered over with houses built for Ute tribe members. And that's where things begin connecting to our story.

"The one right at Turnkey, there is supposed be a graveyard right in there," a former tribal police officer told us in an interview, "but still they built houses. When they built apartments, they built them right over top of that graveyard. Black soldiers, mostly black soldiers in there . . . My grandmother told me about that years ago."

The tale reeks of irony. After decades of spooky Hollywood stories about greedy Caucasians building housing developments over Indian burial grounds, thus unleashing hostile Native American poltergeists bent on revenge, is it possible that Indian opportunists may have disturbed the spirits of dead African-American soldiers who, in life, were steeped in mystical arts? By building homes over a known graveyard, did the Utes awaken an unknown force that has since plagued them with ongoing appearances by unearthly beasts and other inexplicable phenomena? It's a question tinged with superstition and sensationalism, but it's one that some tribe members have asked themselves for years.

When Congress designated the seven thousand acres of Ute land as part of the "public domain," it inadvertently exempted that acreage from any official control or law enforcement. A bawdy collection of saloons and brothels quickly sprang up, and the soldiers of Fort Duchesne were among the most loyal customers, even though the so-called Duchesne Strip was officially off-limits to all military personnel. Outlaws like Butch Cassidy and Elzy Lay hid out in the strip since no lawmen had jurisdiction there. Soldiers, both black and white, drank and cavorted with miners, with outlaws, with prostitutes, and with Utes. As the soldiers returned to the fort following their drunken forays, they often passed by a ravine that became a handy spot for disposing of empty whiskey bottles.

So many bottles were tossed into the ravine that it eventually earned the name Bottle Hollow, a term still used today. Bottle Hollow is now mostly covered by water. In 1970, the federal government authorized the creation of a reservoir on Ute land in partial repayment for the diversion of tribal waters for the Central Utah Water Project. Today the Bottle Hollow reservoir covers some 420 acres and is a popular fishing spot. The fact that Bottle Hollow almost directly abuts Skinwalker Ranch is not lost on the Utes or other local residents.

The reservoir has a mysterious legacy of its own, one that seems inextricably linked to the ranch. The Utes have long believed that Bottle Hollow is inhabited by one or more large aquatic snakes, something akin to the sea serpent legends that are attached to other, much older bodies of water around the world. Eyewitness reports of serpent sightings in the reservoir date back almost to the time when Bottle Hollow was first filled with water. Obviously, the reservoir isn't old enough to be inhabited by a Paleolithic oddity that somehow survived into modern times. But what are we to make of the statements made to us by several seemingly honest witnesses, people who didn't want any public attention whatsoever?

One eyewitness is the same tribal police officer who told us about the graveyard of the Buffalo Soldiers. "We used to see things crawling around in the water that looked like giant snakes," he told us. "[It] would swim straight down from the marina and go all the way down to the bottom end. You could see it on moonlit nights. I seen that, well, everybody, the other guys have seen that snake in there too."

Tribal police officers say an inordinate number of drowning cases have occurred in Bottle Hollow over the years, and at least some of them are unofficially attributed to the presence of the mystery snake. One case that was investigated by police involved the death of a Ute woman who was swimming at night with a male companion. Witnesses on the beach said the woman screamed that something in the water had grabbed her and was pulling her under. Her companion told officers that he dove under the water and grappled with a huge snake in an effort to free the woman, but she was dead by the time he got her back to the surface. Obviously, there are other potential explanations for what occurred that night, but the witnesses on the beach supported this version of the event and investigators took the report seriously.

There are also numerous accounts of strange lights that have been seen entering and leaving the waters of Bottle Hollow. In 1998, a police officer told us that he saw a "large light" plunge into the middle of the reservoir and then quickly exit before flying away into the night sky. The witness did not remember whether the object made any kind of splashing sound during its entry or emergence from the dark water. In 2002, we interviewed four young Caucasian men who said they had recently been on the beach with their dates when a blue-white ball of light flew out of the darkness from the direction of the Gorman ranch. The glowing ball dove into the water just a few feet from the shore, then emerged seconds later. The mystery object had changed its shape while submerged, the witnesses said. From the original ball shape that entered the water, it emerged as something resembling a shimmering, maneuverable belt-shaped shaft of light. After performing a brief writhing aerial dance, the belt of light zipped away at a high rate of speed, hugging the ground before disappearing below the top of Skinwalker Ridge. After questioning the four men at length about their backgrounds and the sighting itself, we concluded that they were honestly describing the event to the best of their abilities. They certainly weren't seeking publicity and requested that their names not be made public.

One other salient point about the Buffalo Soldiers and the strange occurrences in the vicinity of this graveyard is worth mentioning. A geologic feature now known as Skinwalker Ridge contains an unusual carving, which was discovered by Tom Gorman and was later examined by the scientific team that would investigate the ranch. It's an inscription located several feet below the top of the ridge, as if someone had hung suspended from the rim in order to carve it into the rock. The best guess is that the carving is one hundred or more years old. It's a Masonic symbol, no question about it, a strong suggestion that at least a few of the Freemasons among the Buffalo Soldiers may have visited the Gorman Ranch a long time ago.

Today, Uintah County, in the heart of the basin, consists of forty-four hundred square miles and more than twenty-five thousand residents, most of them white Mormons. Nearly a tenth of the population is Ute Indian, with their own settlements near White Rocks, Fort Duchesne, and Randlett. The county's economy is based mostly on ranching and farming, with cattle, hay, and alfalfa the principal agricultural products. Geologists describe the region as "an unusually rich source of hydrocarbon-bearing materials," including oil, oil shale, tar sands, and coal. Oil and gas deposits were discovered in the area in the 1950s and are still being exploited. It is not uncommon to see oil rigs pumping away on almost any main road in the region. In addition, Uintah County officials often refer to their area as "Dinosaur Country" because of the prevalence of fossilized remains from prehistoric times.

Of course, the Gormans were completely unaware of this history of the area when they bought their ranch in the Uinta Basin. They had escaped the frying pan of small-town gossip and stepped into the terrifying fire of the unknown.

CHAPTER 4

The Weirdness

The Gormans tried to ignore the wolf incident. It had been too surreal. One of their new neighbors remarked casually to them a few days later that their land was home to a herd of large wolves. They were relieved to hear that their family was not party to some collective delusion. Other people had seen the wolves too.

A few weeks later, Ellen was driving in her gray Chevette back to their homestead. She was coming home from her new job with the local mortgage company. She had opened the gate to the property and closed it behind her. As she sat in the car, she noticed a movement to her left in her peripheral vision. She gasped. The wolf was huge, and it had silently approached within thirty feet of her. Now it stood outside her window. As she stared into the friendly light-blue eyes of the huge animal, she felt a knot of fear tighten. The animal's head stood over the roof of her car. This was no ordinary wolf—it resembled the bullet-proof animal they had encountered only a few weeks previously.

Easily visible in the gathering twilight was another animal, all black. It stood farther away from the car and appeared more reserved. It was large, but not quite as big as the wolf. It looked like a very weird dog but unlike any she had ever seen. Perhaps it was an exotic breed created by centuries of mating on the nearby Ute Indian reservation? Maybe one of her Ute neighbors owned this wolf? The dog's head was much too large for its body, but the body was still big. Now thoroughly alarmed, Ellen slammed the gas pedal with her foot and drove quickly along the last half mile to the homestead. She made a mental note to complain to the local tribal office in Fort Duchesne the following day. No matter how placid and tame these huge wolves appeared to be, they were still wild animals. She was determined that they would not be allowed to roam freely on the property, especially since the family was preparing to move their extremely expensive herd of registered animals onto the grass fields within a few more weeks. Winter was approaching.

Ellen was really puzzled the next day when her polite requests to rein in the wolves were met with blank stares and uncomprehending silence at the tribal office. Nobody owned any wolves around here, she was told. In fact, wolves had not been seen in this part of Utah for seventy years: the last wolf in Utah had been shot in 1929. (These events happened a full year before a herd of gray wolves was transplanted into Yellowstone Park and into Central Idaho in 1995, and in December 2002 one of those wolves would be caught in a trap near Ogden, Utah. In 2004, a couple of wolves were spotted around Vernal.) The soft-spoken tribal official eventually told Ellen that she must be mistaken. Perhaps she had seen a coyote?

Ellen was furious. She left the tribal office convinced that somebody was not telling the truth. She remembered that the huge animal had to bend down to breathe on her car window. It was about four times the size of any wolf she had ever heard about. And the disquieting incident of the bulletproof wolf came back to haunt the family. They saw the huge animals in the distance a couple more times in the next few weeks. Then they seemed to disappear from the face of the Earth. The family became so busy with ranch chores that Ellen quickly forgot about the huge animals; out of sight was definitely out of mind.

Ellen enjoyed their tranquil countryside far away from any snoopy neighbors. Every evening after finishing all her chores, she liked nothing better than to walk on the property, when she could look at the stars in a pristine dark sky. There was no pollution in this part of Utah. Her sharp eyes were able to pick up the green-red star Betelgeuse when it was visible in the night sky. She loved the silence.

In late 1994, on her second walk along the top of the two-hundred-foot ridge, she was humming joyfully to herself when she felt a huge rush of wind as something very big flew past her, just missing her. She ducked instinctively. Whatever it was, it had missed her by a few feet. Whatever had flown past her had created significant turbulence. It must have been pretty big, she thought to herself. There was no other noise apart from that strange rush of wind.

A bat? It was much too big and too fast for a bat. This thing was very solid and bulky, and it was flying fast. Five minutes later, as she resumed her brisk walk, it happened again. This time it was even closer. Again she ducked. She should have seen some kind of silhouette, because this time the thing was flying west directly toward the light of the setting sun. There was enough light still in the sunset to spot something that big. But she saw nothing. She felt a twinge of uneasiness. Could it have been a bird? A lot of birds don't fly in the darkness. She was puzzled. She turned around to go back. This was getting creepy. She breathed a sigh of relief when she got back to her family. She decided not to tell them.

During the days when Tom was out working on the ranch and the kids were at school, Ellen began to seriously question her memory. She would leave a kitchen utensil on the counter and go outside for a moment and return to find the utensil missing. It would later turn up somewhere unexpected. She couldn't figure it out. She began to suspect that one of her kids was playing games with her. Similar oddities happened a couple of times per week.

Ellen was getting increasingly worried about her memory when one evening Tom stormed in and demanded to know who had hidden the post digger. He was irate. Both kids looked up from their homework. They were puzzled. Ellen told him that all three had been in the house for the past couple of hours. Tom said that he had left the seventy-pound implement on the ground and gone to his truck to get a wrench, returned a couple of minutes later, and it was gone. He was angry because he wanted to finish mending a broken fence, and it was almost dark. The family went outside to help him look. Thirty minutes later they gave up. Tom was silent and frustrated the whole evening.

Two evenings later he rushed in and demanded to know who had taken his pliers. He had left them on a fence post, turned around, and when he turned back, they were gone. He was seething. Again the family looked for the pliers, but to no avail. It was then that Ellen decided to fess up. The same thing had been happening to her, she explained. She reeled off multiple instances. The family stared at her as she spoke. Tom grew quiet. The look on his face told her everything. He had begun to suspect that something was terribly wrong. Shortly afterward he went outside to think. She knew better than to come out to talk to him. This was not the time for a family conference. There was too much to do.

Tom had finally moved his entire herd of high-end, registered breeding Black Angus and Simmental cattle onto the new ranch.

Around the same time, Dave, Tom's nephew, had arrived to visit with the family for a few weeks. Dave was not exactly the outdoor type, having lived most of his life in the city. Tom was determined to break him into the ranching lifestyle during his stay. One night, he told Tad and Dave to accompany him on an outdoor jaunt. Tom knew that Dave was apprehensive of the dark, so he wanted to break the youngster of this fear. The three began walking casually through the property to check on the cows. It was a beautiful dusk, and Tom was appreciating the lovely clear sky that still held some light. A few hundred feet to the north, the ridge was getting darker by the minute.

A flash of annoyance hit Tom as he spied the lights of a trespassing RV about half a mile to their west. He had little patience for trespassers who ignored private property and hunted on other people's land. He had let it go a few times before when he had seen distant lights on his property, but this time he was going to tell these louts off. He pointed out the RV to the two teenagers, and the three of them increased their pace. When they were about two hundred yards away, the RV started moving away from them. Tom was momentarily puzzled. How could it have seen them? Perhaps the trespassers had night-vision equipment, Gorman thought to himself. He and the boys broke into an easy jog. He did not want this idiot to start breaking fence lines as the RV tried to escape. The headlamp in front and the red light behind were moving very smoothly now. Tom wondered why the vehicle was not bouncing over the ruts.

Suddenly the lights from the object seemed to rise a few feet from the ground. Gorman's brow puckered. "What's goin' on?" Tad muttered. They were covering the ground quickly now, trying to catch the RV. They could see that it had gradually increased its pace as it maintained the same distance from them. All three were now running, and again they could see the lights moving a few feet off the ground. As they came to one of the fences through their property, it dawned on Gorman what was happening. The thing was somehow lifting itself over the fence lines! It had already gone over a couple with apparent ease. This was when he felt the first chill. How could an RV be climbing over fences?

The chase continued and Tom was breathing heavily. They had now entered the last pasture before the very end of his property, and that pasture was bound on the western end by a line of Russian olives placed thickly together and right behind a stout five-foot-high barbed-wire fence. Tom grunted with satisfaction as he ran. The bastards were trapped.

He still could not hear the vehicle's engine and he wondered why. They were running hard in the darkness now, and the red taillight of the vehicle was still about two hundred yards in front. Tom kept waiting for the object to slow down as it neared the impenetrable barrier that formed the western limit of his property. The boys were about ten yards in front of him and he was gasping for air now. But since they were only moments from catching these intruders, Gorman kept running. He kept glancing down at the rough rutted terrain as he ran, making sure that any obstacles were not going to trip him up.

Suddenly a loud gasp from the boys made him look up. The RV was now definitely in the air. All three stopped to watch. With the red light on its tail, it climbed smoothly, slowly, and silently toward the top of the tree line. Those trees were more than fifty feet high. As the object crested the tree line, the bewildered trio saw the shape of the vehicle perfectly silhouetted against the horizon. It was no RV. The object was roughly oblong, shaped like a large refrigerator, with a headlight in front and a red light behind. All three watched in complete silence as the object slowly disappeared over the trees in the distance. It was flying smoothly and slowly, almost casually. There was no sound.

Tom's breathing was still coming in painful gulps. He felt cold chills running through his body even as he sweated profusely. The boys, still gazing openmouthed, turned to Gorman looking for an explanation. Even from several feet away, he could see that Dave was crying softly in the darkness. The fourteen-year-old was obviously very frightened and was deeply disturbed by something this bizarre. Tom knew that his son was made of sterner stuff, but Tad too was waiting for an explanation. "I have no idea," Tom muttered as he turned away, trying to figure out exactly what had happened. What was really spooky was that they never heard the sound of an engine, even from 150 yards away.

What will I tell Ellen? he wondered as they trudged back. His mind was racing, looking for some kind of rational explanation for what they had seen. Some kind of military exercise out in the wilds of Utah? It made no sense to him. Why would they test exotic hardware on someone's private property? Nothing added up. They walked in silence all the way back to the homestead, the young fourteen-year-old shivering with fear. Dave's stay would be cut short. His parents let the family know that the youngster would not return to his cousins while they still lived on that property in Utah.

A few weeks later, Tom and Ellen were walking out on the trail heading west and enjoying the cool air about an hour after sunset. The trail passed close to the ridge that marked the northern boundary of the Gorman property. The bluff was composed of dried mud and sandstone, and was still noticeably red as darkness intensified.

Tonight, there was no sign of rain, and the Gormans chatted quietly as they walked. Suddenly a loud metallic sound came from their right, cutting through the nighttime stillness. Startled, they stopped abruptly. Both had acute hearing. A few seconds later, they heard the noise again. It sounded like metal being banged on metal, and it seemed to be coming from about a hundred feet above them in the darkness. Tom was puzzled. What could be making the noise?

Ellen clutched his sleeve and silently pointed in front of them. Tom saw the bright light about a hundred yards out. It was a vehicle. "Probably some miners lost," he said, putting two and two together. They felt slight apprehension as they walked toward it. As the couple approached the object, it lifted off the ground, moved about fifty yards away, and slowly settled back to the ground. "It's that thing Tad and I saw a while back," Tom muttered to his wife.

Slowly they walked in the direction of the bright white beam that stabbed the darkness in front of it. They could just make out the refrigerator shape behind the light. And the reddish glow behind the object was familiar to Tom.

As they drew nearer, the object again lifted off the ground and glided smoothly away in total silence. Again they tried to gain ground, but each time the object repeated the frustrating maneuver. It was clearly watching and reacting to them, trying to keep them at arm's length. Beneath his nervousness, Tom felt a spark of outrage. Who the hell do they think they are? he thought angrily.

In the distance behind them, they both heard the mysterious metallic sound again. They turned away from the chase momentarily to try to get a fix on the sound. When they turned back, there was no sign of the light in front of them. The vehicle had either turned off its lights or had vanished. Slowly they walked toward the spot where they had last seen it. Nothing stirred in the still night. They passed the spot on the path where it had been, but there were no tracks in the hard-baked mud.

They didn't talk much after that, although they walked the entire length of the ranch to the western boundary. The night appeared to have swallowed the mysterious flying refrigerator. Tom knew then for certain that something very unusual was unfolding on the ranch they had purchased just a few months before.

It was around this time that the Gorman family first heard the rumors that were circulating in town. The scuttlebutt being whispered in local coffee shops and stores acknowledged the legends told by the Utes about the secret history of the Gorman property. In short, the ranch was considered to be off-limits for tribe members. Piece by piece, the Gormans learned that their land was cursed and no Native American would ever set foot on it. Why, the Gormans wondered. Nobody would give them a straight answer.

CHAPTER 5
The Curse

In the religion and cultural lore of Southwestern tribes, there are witches known as skinwalkers who can alter their shapes at will to assume the characteristics of certain animals. Most of the world's cultures have their own shapeshifter legends. The best known is the werewolf, popularized by dozens of Hollywood movies. European legends as far back as the 1500s tell stories about werewolves. (The modern psychiatric term for humans who believe they are wolves is *lycanthropy.*) The people of India have a weretiger legend. Africans have stories of wereleopards and werejackals. Egyptians tell of werehyenas.

In the American Southwest, the Navajo, Hopi, Utes, and other tribes each have their own version of the skinwalker story, but basically they boil down to the same thing—a malevolent witch capable of being transformed into a wolf, coyote, bear, bird, or any other animal. The witch might wear the hide or skin of the animal identity it wants to assume, and when the transformation is complete, the human witch inherits the speed, strength, or cunning of the animal whose shape it has taken.

"The Navajo skinwalkers use mind control to make their victims do things to hurt themselves and even end their lives," writes Doug Hickman, a New Mexico educator. "The skinwalker is a very powerful witch. [It] can run faster than a car and can jump mesa cliffs without any effort at all."

For the Navajo and other tribes of the Southwest, the tales of skinwalkers are not mere legend. Just ask Michael Stuhff. A Nevada attorney, Stuhff is likely one of the few lawyers in the history of American jurisprudence to file legal papers against a Navajo witch. He has often represented Native Americans in his practice. He understands Indian law and has earned the trust of his Native American clients, in large part because he knows and respects tribal religious beliefs.

As a young attorney in the mid-1970s, Stuhff worked in a legal aid program based near Genado, Arizona. Many, if not most, of his clients were Navajo. His legal confrontation with a witch occurred in a dispute over child custody and financial support. His client, a Navajo woman who lived on the reservation with her son, was asking for full custody rights and back child support payments from her estranged husband, an Apache. At one point during the legal wrangling, the husband got permission to take the son out for an evening but didn't return the boy until the next day. The son later told his mother what had transpired that night.

According to the son, he spent the night with his father and a "medicine man." They built a fire atop a cliff and, for many hours, the medicine man performed ceremonies, songs, and incantations around the fire. As dawn broke, the three traveled into a wooded area near a cemetery, where they dug a hole. Into the hole, the medicine man deposited two dolls. One of the dolls was made of dark wood, the other of light wood. It was as if the two dolls were meant to represent the mother and her lawyer. Although Stuhff wasn't sure how seriously to take the news, he recognized that it certainly didn't sound good, so he sought the advice of a Navajo professor at a nearby community college.

"He told me that the ceremony I had described was very powerful and very serious, and that it meant that I was supposed to end up buried in that cemetery," Stuhff says. "He also said that a witch can perform this type of ceremony only four times in his life, because if he tries it more than that, the curse would come back on the witch himself. He also told me that if the intended victim found out about it, then the curse would come back onto the person who had requested it."

Stuhff thought about a way to let the husband know that he had found out about the ceremony, so he filed court papers that requested an injunction against the husband and the unknown medicine man, whom he described in the court documents as "John Doe, A Witch." The motion described in great detail the alleged ceremony. The opposing attorney appeared extremely upset by the motion, as did the husband and the presiding judge. The opposing lawyer argued to the court that the medicine man had performed "a blessing way ceremony," not a curse. But Stuhff knew that the judge, who was a Navajo, could distinguish between a blessing ceremony, which takes place in Navajo hogans (homes), and what was obviously a darker ceremony involving lookalike dolls that took place in the woods near a cemetery. The judge nodded in agreement when Stuhff responded. Before the judge could rule, Stuhff requested a recess so that the significance of his legal motion could sink in. The next day, the husband capitulated by agreeing to grant total custody to the mother and to pay all back child support.

"I took it very seriously because *he* took it seriously," Stuhff says. "I learned early on that sometimes witches will do things themselves to assist the supernatural, and I knew what that might mean."

Whether or not Stuhff literally believes that witches have supernatural powers, he acknowledges that this belief is strongly held in the Navajo nation. Certain communities on the reservation had reputations as witchcraft strongholds, he says. It is also not known whether the witch he faced was a skinwalker or not. "Not all witches are skinwalkers," he says, "but all skinwalkers are witches. And skinwalkers are at the top. They are a witch's witch, so to speak."

According to University of Nevada–Las Vegas anthropologist Dan Benyshek, who specializes in the study of Native Americans of the Southwest, "Skinwalkers are purely evil in intent. I'm no expert on it, but the general view is that skinwalkers do all sorts of terrible things—they make people sick, they commit murders. They are grave robbers and necrophiliacs. They are greedy and evil people who must kill a sibling or other relative to be initiated as a skinwalker. They supposedly can turn into wereanimals and can travel in supernatural ways."

Benyshek and other scientists do not necessarily endorse the legitimacy of the legends, but they recognize the importance of studying stories about skinwalkers because the power of the belief among Native Americans manifests itself in ways that are very real. "Oh, absolutely," says Benyshek. "Anthropologists have conducted scientific investigations into the beliefs in Native American witchcraft because of the effects of such beliefs on human health."

Anthropologist David Zimmerman of the Navajo Nation Historic Preservation Department explains: "Skinwalkers are folks that possess knowledge of medicine, medicine both practical (heal the sick) and spiritual (maintain harmony), and they are both wrapped together in ways that are nearly impossible to untangle."

As Zimmerman suggests, the flip side of the skinwalker coin is the power of tribal medicine men. Among the Navajo, for instance, medicine men train over a period of many years to become full-fledged practitioners in the mystical rituals of the Diné (Navajo) people. The U.S. Public Health Service now works side by side with Navajo medicine men because the results of this collaboration have been proved, time and again, in clinical studies. The medicine men have shown themselves to be effective in treating a range of ailments.

"There has been a lot of serious research into medicine men and traditional healers," says Benyshek. "As healers, they are regarded as being very effective in some areas."

But there is a dark side to the learning of the medicine men. Witches follow some of the same training and obtain similar knowledge as their more benevolent colleagues, but they supplement both with their pursuit of the dark arts, or black magic. By Navajo law, a known witch has forfeited its status as a human and can be killed at will. The assumption is that a witch, by definition, is evil.

"Witchcraft was always an accepted, if not widely acknowledged part of Navajo culture," writes journalist A. Lynn Allison. "And the killing of witches was historically as much accepted among the Navajo as among the Europeans." Allison has studied what she calls the "Navajo Witch Purge of 1878" and has written a book on the subject. In that year, more than forty Navajo witches were killed or "purged" by tribe members because the Navajo had endured a horrendous forced march at the hands of the U.S. Army in which hundreds were starved, murdered, or left to die. At the end of the march, the Navajo were confined to a bleak reservation that left them destitute and starving. The gross injustice of their situation led them to conclude that witches might be responsible, so they purged their ranks of suspected witches as a means of restoring harmony and balance. Tribe members reportedly found a collection of witch artifacts wrapped in a copy of the Treaty of 1868 and "buried in the belly of a dead person." It was all the proof they needed to unleash their deadly purge.

"Unexplained sickness or death of tribal members or their livestock could arouse suspicion of witchcraft," Allison writes in her book. "So could an unexplained reversal of fortune, good or bad."

In the Navajo world, where witchcraft is important, where daily behavior is patterned to avoid it, prevent it, and cure it, there are as many words for its various forms as there are words for various kinds of snow among the Eskimos.

The Navajo people do not openly talk about skinwalkers, certainly not to outsiders. Author Tony Hillerman, who has lived for many years among the Navajo, used the skinwalker legend as the backdrop for one of his immensely popular detective novels, one that pitted his intrepid Navajo lawmen Jim Chee and Joe Leaphorn against the dark powers of witchcraft. The following excerpt is from *Skinwalkers:*

> "You think that if I confess that I witched your baby, then the baby will get well and pretty soon I will die," Chee said. "Is that right? Or if you kill me, then the witching will go away."
> "You should confess," the woman said. "You should say you did it. Otherwise, I will kill you."

Hillerman has been harshly criticized by some Navajo for bringing unwanted attention to the subject of skinwalkers. "No one who has ever lived in the Navajo country would ever make light of this sinister situation," wrote one critic after Hillerman's book was dramatized on PBS in 2003.

Anthropologist Zimmerman explains why so little information is available on skinwalkers: "Part of the reason you won't find a lot of information about skinwalkers in the literature is because it is a sensitive topic among the Diné. This is often referred to as proprietary information, meaning it belongs to the Diné people and is not to be shared with the non-Diné."

We know from experience that is it extremely difficult to get Native Americans to discuss skinwalkers, even in the most general terms. Practitioners of *adishgash*, or witchcraft, are considered to be a very real presence in the Navajo world. Few Navajo want to cross paths with *naagloshii* (or *yee naaldooshi)*, otherwise known as a skinwalker. The cautious Navajo will not speak openly about skinwalkers, especially with strangers, because to do so might invite the attention of a witch. After all, a stranger who asks questions about skinwalkers just might be one himself, looking for his next victim.

"They curse people and cause great suffering and death," one Navajo writer explained. "At night, their eyes glow red like hot coals. It is said that if you see the face of a naagloshii, they have to kill you. If you see one and know who it is, they will die. If you see them and you don't know them, they have to kill you to keep you from finding out who they are. They use a mixture that some call corpse powder, which they blow into your face. Your tongue turns black and you go into convulsions and you eventually die. They are known to use evil spirits in their ceremonies. The Diné have learned ways to protect themselves against this evil and one has to always be on guard."

One story told on the Navajo reservation in Arizona concerns a woman who delivered newspapers in the early morning hours. She claims that, during her rounds, she heard a scratching on the passenger door of her vehicle. Her baby was in the car seat next to her. The door flung open, and she saw the horrifying form of a creature she described as half man, half beast, with glowing red eyes and a gnarly arm that was reaching for her child. She fought it off, managed to pull the door closed, then pounded the gas pedal and sped off. To her horror, she says, the creature ran along with the car and continued to try to open the door. It stayed with her until she screeched up to an all-night convenience store. She ran inside, screaming and hysterical, but when the store employee dashed outside, the being had vanished. Outsiders may view the story skeptically, and any number of alternative explanations might be suggested, but it is taken seriously on the Navajo reservation.

Although skinwalkers are generally believed to prey only on Native Americans, there are recent reports from Anglos claiming they encountered skinwalkers while driving on or near tribal lands. One New Mexico Highway Patrol officer told us that while patrolling a stretch of highway south of Gallup, New Mexico, he had had two separate encounters with a ghastly creature that seemingly attached itself to the door of his vehicle. During the first encounter, the veteran law enforcement officer said the unearthly being appeared to be wearing a ghostly mask as it kept pace with his patrol car. To his horror, he realized that the ghoulish specter wasn't attached to his door after all. Instead, he said, it was running alongside his vehicle as he cruised down the highway at a high rate of speed.

The officer said he had a nearly identical experience in the same area a few days later. He was shaken to his core by these encounters but didn't realize that he would soon get some confirmation that what he had seen was real. While having coffee with a fellow highway patrolman not long after the second incident, the cop cautiously described his twin experiences. To his amazement, the second officer admitted having his own encounter with a white-masked ghoul, a being that appeared out of nowhere and then somehow kept pace with his cruiser as he sped across the desert. The first officer told us that he still patrols the same stretch of highway and that he is petrified every time he enters the area.

One Caucasian family still speaks in hushed tones about its encounter with a skinwalker, even though it happened in 1983. While driving at night along Route 163 through the massive Navajo reservation, the four members of the family felt that someone was following them. As their truck slowed to round a sharp bend, the atmosphere changed, and time itself seemed to slow down. Then something leaped out of a roadside ditch at the vehicle.

"It was black and hairy and was eye level with the cab," one of the witnesses recalled. "Whatever this thing was, it wore a man's clothes. It had on a white and blue checked shirt and long pants. Its arms were raised over its head, almost touching the top of the cab. It looked like a hairy man or a hairy animal in man's clothing, but it didn't look like an ape or anything like that. Its eyes were yellow and its mouth was open."

The father, described as a fearless man who served two tours in Vietnam, turned completely white, the blood drained from his face. The hair on his neck and arms stood straight up, like a cat under duress, and noticeable goose bumps erupted from his skin. Although time seemed frozen during this bizarre interlude, the truck continued on its way, and the family was soon miles down the highway.

A few days later, at their home in Flagstaff, the family awoke to the sounds of loud drumming. As they peered out their windows, they saw the dark forms of three "men" outside their fence. The shadowy beings tried to climb the fence to enter the yard but seemed inexplicably unable to cross onto the property. Frustrated by their failed entry, the men began to chant in the darkness as the terrified family huddled inside the house.

The story leaves several questions unanswered. If the beings were skinwalkers, and if skinwalkers can assume animal form or even fly, it isn't clear why they couldn't scale a fence. It is also not known whether the family called the police about the attempted intrusion by strangers.

The daughter, Frances, says she contacted a friend, a Navajo woman who is knowledgeable about witchcraft. The woman visited the home, inspected the grounds, and offered her opinion that the intruders had been skinwalkers who were drawn by the family's "power" and that they had intended to take that power by whatever means necessary. She surmised that the intrusion failed because something was protecting the family, while admitting that it was all highly unusual since skinwalkers rarely bother non-Indians. The Navajo woman performed a blessing ceremony at the home. Whether the ceremony had any legitimacy or not, the family felt better for it and has had no similar experiences since then.

This disturbing account is not offered as proof of anything, particularly since we have not personally interviewed the witnesses. It is presented only as an illustration of the intense fear and unsettling descriptions that permeate skinwalker lore, which is accepted at face value by the Native Americans for whom the skinwalker topic is not just a spooky children's story.

So exactly how and when did the skinwalker legend intersect with the Gorman ranch? Junior Hicks says his friends in the Ute tribe believe that the skinwalker presence in the Uinta Basin extends back at least fifteen generations. The Utes, described by historians as a fierce and warlike people, were sometimes aligned with the Navajo against common enemies during the 1800s. But the alliance didn't last. When the Utes first acquired horses from the Spanish, they enthusiastically embraced the Spanish example by engaging in the slave trade. They reportedly abducted Navajos and other Indians and sold them in New Mexico slave markets. Later, during the American Civil War, some Ute bands took orders from Kit Carson in a military campaign against the Navajo. According to Hicks, the Utes believe that the Navajo put a curse on their tribe in retribution for many perceived transgressions. And ever since that time, Hicks was told, the skinwalker has plagued the Ute people.

The ranch property has been declared off-limits to tribal members because it lies in the path of the skinwalker. Even today, Utes refuse to set foot on what they see as cursed land. But the tribe doesn't necessarily believe that the skinwalker lives on the ranch. Hicks says the Utes told him that the skinwalker lives in a place called Dark Canyon, which is not far from the

ranch. In the early 1980s, Hicks sought permission from tribal elders to explore the canyon. He'd been told there are centuries-
old petroglyphs in Dark Canyon, some of which depict the
skinwalker. But the tribal council denied his request to explore
the canyon. One member later confided to Hicks that the tribe
denied the request because it did not want to disturb the skinwalker for fear that it might "create problems." The tribe's advice
to Hicks: "Leave it alone."

Dan Benyshek suggests that some parts of this Utes' account
don't add up. He thinks it unlikely that the Navajo would enlist
the assistance of a skinwalker to carry out their revenge on the
Utes, no matter how much the tribe might want some payback
on their enemy. "The skinwalkers are regarded as selfish, greedy,
and untrustworthy," Benyshek says. "If the Navajo knew someone to be a skinwalker, they would probably kill him, not ask for
his help with the Utes. Besides, even if he was asked, the skinwalker would be unlikely to help the Navajo get revenge, since
his motives are entirely evil and self-serving. From the Navajo
perspective, this story doesn't make sense."

But from the Ute perspective, it could ring true. "The Utes
could very likely have concluded that the curse is real," explains
Benyshek. "Different tribes or bands would often tell stories
about the evil motives of other tribes they were in conflict with,
about how another tribe was in league with witches, or how
other tribes were cannibals. The Utes might tell themselves this
story as a way to explain their own misfortunes."

Hicks told us that the Indians say they see them a lot. "When
they go out camping," he says, "they sprinkle bark around their
campsites and light it as protection against these things. But it's
not just Indians. Whites see them, too." Like his Ute neighbors,
Hicks sometimes uses the terms *skinwalker* and *Sasquatch* inter-
changeably. He says he's seen photographs of the telltale huge
footprints often associated with Bigfoot, taken in the vicinity of
the Gorman ranch. But whether it was a run-of-the-mill

Sasquatch or a far more sinister skinwalker isn't always clear, even to those who accept the existence of both.

"There was an incident sixteen years ago where a skinwalker was on a porch in Fort Duchesne," Hicks remembers. "They called the tribal police and tracked it east toward the river. They took some shots at it and thought they hit it because they found blood on the ground, but they never found a body."

We also conducted an interview with a Ute man who worked as a security officer for the tribe. He provided us details about his own encounter with a Bigfoot. Brandon Ware (not his real name) received his police training at an academy associated with the Bureau of Indian Affairs. He says he was working the 10:00 P.M. to 4:00 A.M. shift, guarding a tribal building near Bottle Hollow. Between midnight and 1:00 in the morning, Ware walked up to check on the building and noticed that the guard dogs inside were calm but staring intently through a window at something outside. They weren't barking, he said, just looking.

"I could see this big ol' round thing, you know, in the patio over there," Ware recalls, "and the hair started raising on my neck and I kinda got worried a little bit trying to figure out what things were. I stood there and watched it for a few minutes, then it came over the top and headed down the road. But I could smell it. Even after it was gone, you could smell it."

Ware says that when the creature realized it was being observed, it briefly looked over at Ware, then vaulted over a low wall that surrounded the patio area outside the building. He says it took off running toward the Little Chicago neighborhood, crashing into garbage cans as it moved past the homes, and generating a cacophony of loud barking by every dog in the immediate area. Ware then went into the building and telephoned another on-duty officer who was nearby. By the time Ware left the building, the other officer had pulled up in his patrol car.

Ware told the other officer to turn off his engine so they could listen to the hubbub that was still unfolding among the nearby homes. "We listened a little bit and we could hear it. Then we jumped in and took off. We headed down the hill to see if we could catch up to it."

The two officers didn't see the creature again that night, but they had no trouble tracing its path through the cluster of homes because they were able to follow a noticeable trail of scattered garbage cans. "It must have gone straight on through," Ware recalls. "We could see where cans—people usually tie up their cans—them were all off. I told the other officer, 'Hey man, maybe it picked up them cans and was throwing them at those dogs.'"

Ware provided further details about what he had seen. His initial impression was of something dark and round. But he says that when the creature stood erect to vault over the patio wall, it appeared to be "huge." Ware was carrying a large flashlight at the time of the encounter. He says he was using the flashlight just minutes before while checking the doors of the building, but when he tried to use it to illuminate the creature, the light wouldn't turn on. When the creature took off running down the hill, the flashlight clicked back on.

"He moved quick," he told us. "Whatever it was, it moved. I called him a 'he'—it could have been a she. It could have been whatever, but he moved quick going down through there. But it was kind of cool. It was neat. I never knew it . . . it was something I've never seen before. I've heard about them. I heard the old people talking about some of these things."

Just a few nights later, Ware got a chance for a second look. He and another officer, Bob (not his real name), were patrolling a back road that emerges at a spot known as Shorty's Hill. They emerged from the road into a pasture area that is punctuated by a large rock. "I don't know if it was the same guy or not," Ware says. "It was a big ol' black hairy thing hanging there, and when it turned around, it had big ol' eyes on him. It had big ol' red eyes on him about yea big. We'd just passed it and I told Bob,

'There he is,' and then he come to a screeching halt and we backed up. By the time we got out, it was gone."

Ware described the creature's eyes as being "coal red" and unusually large. He isn't sure whether the headlights of the patrol car might have affected his perception of the beast's eye color, but tends to doubt it. He has no doubt about the presence of the beast itself. "We got out there to go look and we had shotguns and pistols and everything. We were going to blow him away," Ware admits.

When pressed for his opinion of what he had seen, whether it might have been a Sasquatch or even a skinwalker, Ware's response seemed to draw a distinction between the two, but the distinction became blurry as the conversation progressed and Ware explained his understanding of tribal lore.

"Sasquatch, he's an old man, an old man that lives on a mountain," he explained. "He just comes in and looks at people and then he goes back out again. He just lives there all his life, never takes care of himself, and just smells real bad. Almost like, almost like that guy, like he is dirty, dirty human being smell is what it smelled like . . . a real deep, bad odor . . . It smelled like dirty bad underarms . . . The closer I got, the worse the smell got." Could the creature he saw have been a skinwalker?

"Nope," said Ware. "A skinwalker's smaller. A skinwalker is the size of humans, six foot and under. They don't come in most of the time to where the animals are at. They come in where people are at. They can come right here and you'd never know he was standing here looking at you in the middle of the night . . . They can take the shape of anything they want to take the shape of. Like I said, they're medicine."

Ware said that skinwalker sightings among the Utes are not uncommon. He told us of an encounter with two shape-shifters near the Gorman ranch. The figures he described are so unusual, so far outside our own concept of reality as to be almost comical, like something out of a Saturday morning cartoon. One local who saw them in the road in Fort Duchesne described them as humans with dog heads smoking cigarettes. But Ware was perfectly serious in his description. He certainly did not bare his soul for comic effect, and we have no interest in making light of his story. For him, and for many others, skinwalkers are as real as the morning sun or the evening moon. They are a part of everyday life, and they most certainly are integral to the story of the Gorman ranch.

Could the Utes have used the skinwalker curse as an all-encompassing explanation for their assorted tribal misfortunes, as Benyshek asks? Or are they relying on the legend as an umbrella explanation for the wide range of paranormal events that have been reported in the vicinity of their lands for generations—in particular, in the vicinity of the ranch?

If a skinwalker really is a shape-shifter, capable of mind control and other trickery, might it also have the ability to conjure up nightmarish visions of Bigfoot or UFOs? Could it steal and mutilate cattle, incinerate dogs, generate images of monsters, unknown creatures, or extinct species, and could it also frighten hapless residents with poltergeistlike activity?

At the very least, the skinwalker legend might be a convenient way for the Utes to grasp a vast menu of otherwise inexplicable events, the same sort of events that might stymie and confuse a team of modern scientists.

CHAPTER 6
High Strangeness

Winter had set in and the temperatures dropped precipitously. Thirty or forty below was not unusual for this time of the year in northeastern Utah. This was very different from the mild winters in New Mexico the family had left behind, but they did not regret their move. At least, not yet.

Tom had taken to spending a lot of time out at night trying to get a handle on the weirdness on his property, which had really begun to escalate once he brought his cattle onto the acreage. Every night, Gorman crept like a ghostly wraith around the land, trying to catch the intruders in a better light. He was getting nowhere fast. They seemed to have very good antennae. He could see them only in the distance, never up close.

At the end of one fruitless night, the hard-packed snow crunched underfoot as he walked slowly back from the western part of the ranch. It was probably thirty below and Tom wanted to call it a night. He was freezing cold. As he trudged home, to his north lay the snow-covered slope of the bluff, and the post-blizzard freshness bathed everything in an eerie pale light. A slight movement out of the corner of his eye caught Tom's attention. He turned toward the ridge and his jaw slackened. Starkly outlined against the snow-covered ridge was an aircraft that seemed to have just appeared out of nowhere. It was about thirty or forty feet long, and it reminded Gorman of a snub-nosed, smaller hybrid version of the F-117 and the B-2. But it was completely silent. There was no wind, and the stillness was uncanny.

The aircraft was no more than twenty feet off the ground and, as it moved slightly toward him, a recurring pattern of tiny multicolored lights danced their way across the snow. The lights were obviously coming from the aircraft, but Tom couldn't make out how they were being projected onto the snow. The total silence and barely perceptible movement of the object convinced Tom that there was no way he would have seen it if it hadn't been so plainly silhouetted against the white snow. Its jet-black color and the noiseless slow motion combined to give him the impression that it was defying gravity.

Tom crouched in the snow to avoid being seen. It was moving slowly along the ridge and seemed to be quartering the ground as if looking for something. Its tiny "disco" lights traced their silent pattern in the snow as the object slowly turned at the end of the ridge and then resumed its journey over the snow, searching for God knows what. It was a slow, methodical search, but for what? The outline of the object, less than a hundred yards away, gave Tom a good view of its short, matte-black wings. It was definitely like a small version of the F-117 "stealth fighter." The odd angled design seemed quite similar. But Tom knew that the F-117 was extremely noisy.

As Tom stretched his aching, freezing body, his bones cracked in the still air. Instantly, the disco lights turned off and the object turned toward him. It hovered silently less than a hundred yards away and only fifteen or twenty feet off the ground.

Tom held his breath. The object moved silently away from him without ever increasing its speed and then disappeared into the gloomy night beyond the ridge. Tom heaved a sigh of relief and awkwardly got to his knees in the snow. He looked back several times at the area near the ridge where he had last seen the silent black phantom.

What was this advanced technology? he wondered. Who owned it? And what on earth was it doing there on a ranch in Utah? Gorman silently let himself into the homestead where his family slept and gratefully embraced the warmth.

Several weeks later, Ellen was driving their battered old Chevette home from work when another incident occurred. It was about six o'clock in the evening. Tom was out of town for a couple of days. The kids were staying with friends and she was not looking forward to spending the night alone on this property. She had jumped back into the car after closing the entrance gate behind her when she noticed a black shadow moving slowly over the car. It was like a dark cloud, but the night was clear.

Looking up, she gasped as she saw a large black, triangular object moving slowly and apparently pacing her car as she drove it slowly along the rutted dirt track toward the homestead. She was deathly frightened. The triangular aircraft made no noise as it flew no more than twenty or thirty feet above her car. She could see it out her front windshield and she could see the stubby matte-black wings as she looked out the car's side windows.

She saw tiny multicolored red, green, blue, and yellow lights on the ground on both sides of her as she stepped on the gas. The thing kept pace with her for the short quarter-mile drive, and when she pulled into her driveway, the black triangle continued west over the house until she lost sight of it in the gloomy night. She was now badly frightened. She remembered her husband's description of the silent, stealth shaped—fighter object silhouetted against the snow and she figured the thing was back. She called her husband in his hotel room. He managed to calm her down.

An hour later, she had eaten a quick solitary meal and was washing the dishes. As she looked west out into the dark night, she was startled to see what looked like a large RV parked in the pasture, no more than two hundred yards from her window. The interior of the RV was brightly lit and she could see what looked like a desk inside. Idly, she wondered what on earth an RV was doing trespassing on their property. More to the point, how could it have made it onto the property? After all, there was only one entrance to the ranch and that was right past her home.

Then, a black-colored figure moved into view and sat behind the desk. Her sharp eyesight made out what appeared to be a black uniform, including some kind of headgear. The figure was just sitting at a desk in the night on her property in the middle of a remote ranch in Utah. How weird. The figure suddenly stood up and went to what appeared to be a light-filled doorway. The size of the figure outlined against the brightly lit interior of the RV gave her the chills. The figure looked huge and probably male. If that was a normal RV doorway, he was maybe seven feet tall. And he appeared to be wearing all black clothing, something like a black visor over his face, and knee-high boots.

Suddenly, it occurred to her that the bizarre black triangle she had seen and this threatening black colored silhouetted form might be connected. The figure seemed to be looking out the brightly lit doorway and staring directly at her. She could feel his cold gaze. Quickly, she closed the window drapes and hurriedly called her husband. She wanted him back as quickly as possible. Given the note of high-pitched panic in her voice, Tom decided to return quickly to the ranch. He drove all night and arrived the next morning. Together they walked down to where she had seen the "RV."

Both of them saw the huge footprints in the soft mud at the same time. Ellen became almost hysterical. The size of the boot-shaped prints shocked Tom. They were almost eighteen inches long. The prints did not resemble those of a military boot, as there were no ribs, just a smooth surface with a prominent rounded heel.

This event convinced the Gormans not to let their kids out at night. Something that they couldn't explain was lurking on their property. Ellen used to love walking at night, feeling the wind blowing on her face. That too ended. Henceforth, the Gorman family members were much more cautious. It finally dawned on them that they might be in danger.

CHAPTER 7

Chupas

The Gormans obviously witnessed some unidentified flying objects over their property and felt threatened by them. Was their fear justified? We think it was an appropriate reaction, especially given the fact the rectangular objects they had seen had previously been linked to human injuries—and death. In Brazil, these box-shaped craft that are said to make a sound like a refrigerator are known as *chupas*.

Dr. Jacques Vallee, a scientist who has carefully and intensively investigated the UFO phenomenon for more than four decades, went to South America in the 1980s to investigate the *chupa* wave in Brazil and the numerous reports of UFO close encounters that have resulted in the death or injury of witnesses. The results of his extensive field investigations are detailed in his book *Confrontations: A Scientist's Search for Alien Contact.* For example, in Parnarama in central Brazil, Vallee reports that at least five people had died "following close encounters with what were described as boxlike UFOs equipped with powerful light beams." Many of the victims were hunters who, following the local tradition, had climbed into jungle trees at night to wait for passing animals that they could spotlight and shoot. According to researcher Simon Harvey-Wilson, in an ironic twist, the hunters had themselves been hunted by UFO craft, which injured or killed them using light beams of their own.

Harvey-Wilson continues. "As Vallee reports, these *chupas* (UFOs) 'are said to make a humming sound like a refrigerator or a transformer, and this sound does not change when the object accelerates. The object does not seem large enough to contain a human pilot.' In one case a victim called Dionizio General 'was atop a hill when an object hovered above him and shot a beam in his direction; it was described as "a big ray of fire." The witness, José dos Santos, testified that Dionizio seemed to receive a shock and came rolling down the hill. For the following three days he was insane with terror, then he died.' Witnesses described the light beams as being blinding, like electrical arcs, with pulsating colors inside and smelling unpleasant, which Vallee suspects may be ozone."

By November 1977, according to Vallee, the physician in charge of the health unit on Colares Island, Dr. Wellaide Cecim Carvalho de Oliveira, "had seen no fewer than thirty-five patients claiming injuries related to the chupas. All of them had suffered lesions to the face or the thoracic area." These lesions, which resembled radiation injuries, "began with intense reddening of the skin in the affected area. Later the hair would fall out and the skin would turn black. There was no pain, only a slight warmth. One also noticed small puncture marks in the skin. The victims were men and women of varying ages, without any pattern."

Vallee published a more comprehensive list of symptoms drawn up by Dr. Carvalho:

- a feeling of weakness; some could hardly walk
- dizziness and headaches
- local losses of sensitivity; numbness and trembling
- pallid complexion
- low arterial pressure
- anemia, with low hemoglobin levels
- blackened skin where the light had hit, with several red-purple circles, hot and painful, two to three centimeters in diameter
- two puncture marks inside the red circles resembling mosquito bites, hard to the touch
- hair in the blackened area fell out and did not rejuvenate, as if follicles had been destroyed
- no nausea or diarrhea

In describing their experiences with these light beams, most victims claimed, according to Vallee, that "they were immediately immobilized, as if a heavy weight pushed against their chest. The beam was about [seven or eight centimeters] in diameter and white in color. It never hunted for them but hit them suddenly. When they tried to scream no sound would come out, but their eyes remained open. The beam felt hot, 'almost as hot as a cigarette burn,' barely tolerable. After a few minutes the column of light would slowly retract and disappear." Apart from those who had been killed by these beams, most people's symptoms usually disappeared after seven days.

After asking various forensic pathologists to review his findings, Vallee surmised that "what UFO witnesses describe as 'light' may, in fact, be a complex combination of ionizing and non-ionizing radiation. Many of the injuries described in

Brazil, however, are consistent with the effects of high-power pulsed microwaves." Subsequently he pointed out that pulsed microwaves may "interfere with the central nervous system. Such a beam could cause the dizziness, headaches, paralysis, pricklings, and numbness reported to us by so many witnesses."

Vallee tackled the issue of whether these Brazilian UFOs are deliberately trying to kill people or not. If they are, he considers that they are fairly inefficient. After all, someone in a helicopter with a high-powered rifle and night scope could probably do a better job. He does, however, point out that a radiation beam that was designed merely to stun people at one range might be lethal at another range.

Journalist Bob Pratt also investigated the injuries that occurred during the Brazilian UFO wave and wrote a book about it called *UFO Danger Zone: Terror and Death in Brazil*. Like Vallee, Pratt's investigations of the Brazil wave turned up several deaths and injuries from UFOs. In other words, the New Age bromides that UFOs were actually friendly space brothers intent on the spiritual enlightenment of the human race were not borne out by the experiences of the locals along the northern coast of Brazil.

The issue of injuries from UFOs also cropped up in North America during the last week of 1980. On the evening of December 29, according to accounts by researcher John Schuessler and written accounts by the National Investigations Committee on Aerial Phenomena (NICAP), Betty Cash, Vickie Landrum, and Colby Landrum had visited several small towns in the Piney Woods area of east Texas looking for a bingo game but found that bingo throughout the area had been canceled for Christmas. Instead they decided to have a meal at a restaurant in New Caney. Betty Cash was then a fifty-one-year-old businesswoman who owned a restaurant and a grocery store. Vickie Landrum, fifty-seven, worked for

Betty in the restaurant. Colby Landrum, Vickie's grandson, was seven.

According to the NICAP report, after leaving the restaurant between 8:20 and 8:30 P.M., Betty drove along Highway FM1485, a road usually used only by people who live in the area because it is so isolated. Although the location is only about thirty miles from Houston, it is thinly populated and is covered by oak and pine trees, swamps, and lakes.

At about 9:00, Colby noticed a bright lighted object above the treetops some distance away. He pointed it out to the others. As they drove on, it appeared to get larger and larger. They realized that the object was approaching the road, but they hoped to get by it in time and leave it behind. But before they could do so, the object straddled the road and blocked their way. The object, many times larger than the car, remained hovering at treetop level and sent down an occasional large cone of fire. In between these blasts it would descend until it was no more than about twenty-five feet off the surface of the road. Vickie described it as "like a diamond of fire."

When Betty eventually brought the car to a standstill, the object was only sixty yards away. It looked as if it was made of dull aluminum and glowed so brightly that it lit up the surrounding trees. The four points of the diamond were blunt rather than sharp, and blue spots or lights ringed its center. Had it not come to rest over the road, the cone of fire from its lowest point would have set the forest on fire. The object also emitted an intermittent bleeping sound. The two women got the impression that maybe the giant object was having some kind of engine trouble.

The three of them got out of the car to take a better look at the object. Vickie stood by the open door on the right-hand side of the car with her left hand resting on the car roof. Vickie is a Christian who does not believe in UFOs or extraterrestrial life, and when she saw the bright object she thought the end of the world had come. Because she expected to see Jesus come out of the light, she stared at it intently. Colby begged his grandmother to get back in the car. After about three minutes she did so and told him not to be afraid because "when that big man comes out of the burning cloud, it will be Jesus."

As Vickie held Colby, she screamed at Betty to get back in the car with them. But Betty was so fascinated by the object that she walked around to the front of the car and stood there gazing at it, bathed in the bright light, the heat from it burning her skin. Eventually, as the thing began to move up and away, she moved back to the car door. When she went to open the door, she found the handle so painfully hot that she had to use her leather jacket to protect her hands and get back in the car.

As the three of them watched the departing object, a large number of helicopters appeared overhead. "They seemed to rush in from all directions," Betty recalled later. "It seemed like they were trying to encircle the thing." Within a few seconds the flying diamond had disappeared behind the trees lining the road. It was then that they realized how hot the interior of the car had become. They switched off the heater and put on the air conditioner instead.

When the effects of the bright light had worn off, Betty drove off down the darkened highway. After driving at sixty miles an hour for five minutes, the object was clearly visible some distance ahead and looked like a bright cylinder of light. Even from that distance, it was still lighting up the surrounding area and illuminating the helicopters.

From their new vantage point, they counted twenty-three helicopters. Many of them were identified as the large double-rotor CH-47 Chinooks, the others were very fast single-rotor types and appeared to be Bell-Hueys but were never properly identified. As soon as the flying diamond and helicopters were a safe distance ahead, Betty drove on. When she reached an intersection, she turned away from the flight path of the object and toward Dayton, where the three of them lived. She dropped Vickie and Colby off at their home at about 9:50 and went home by herself. A friend and her children were waiting up for Betty, but by this time she was feeling too ill to tell them about what had happened.

Over the next few hours, Betty's skin turned red as if badly sunburned. Her neck swelled and blisters erupted and broke on her face, scalp, and eyelids. She started to vomit and continued to do so throughout the night. By morning she was almost in a coma. Sometime between midnight and 2 A.M., Vickie and Colby began to suffer similar symptoms, although less severe. At first they suffered a sunburnlike condition and then diarrhea and vomiting.

The following morning, Betty was moved to Vickie's house and all three were cared for there. Betty's condition continued to deteriorate, and three days later she was taken to the hospital. The burns and swelling had so radically altered Betty's appearance that friends who came to visit her in hospital did not recognize her. Her hair began to fall out and her eyes became so swollen that she was unable to see for a week.

Doctors examined all three of them intensively and determined that their conditions suggested that they had been exposed to radiation, possibly ionizing UV or IR. The injuries were not entirely consistent with either ionizing radiation or UV/IR injuries only. Some also suspected that microwave radiation may have literally cooked Betty Cash. Betty died of cancer in 1998, possibly as a result of the radiation she received. Regardless of whether the object they encountered was a secret government project or something else, the injuries they received were real.

The Gormans had every reason to be scared.

CHAPTER 8

The Window

Of all the extraordinary things that occurred at the Gorman ranch, the most common involved the strange, unworldly orange structures that would appear in the western sky. All family members saw these structures dozens of times. They would appear in the sky and seemed to hover low over the cottonwood trees about a mile away. Tom often used a large, four-foot-high tree stump that stood outside the homestead as a vantage point to steady his binoculars or other viewing equipment. His favorite piece of gear was the scope on a night-vision rifle. He could easily hold it steady while leaning on the tree stump and watch the bizarre orange structure about a mile away. Sometimes the object looked flattened and elongated, and sometimes it looked like a large orange setting sun, bigger than a harvest moon and almost perfectly round.

Tom related how one night he set up his scope on the tree stump to look at a gigantic orange object. The detail was astonishing as he looked at the structure hovering silently in the night sky above the row of cottonwoods. Why did it always appear in roughly the same place? The sun had long since set. In the middle of the orange mass Tom could see what looked to him like "another sky." Through the magnifying scope he distinctly saw a blue sky. On this particular night, the orange object looked like it was a window into somewhere else where it was still daylight. Tom felt like it could have been a tear or a rent in the sky about a mile away, and through the rent he could see a different world or perhaps a different time. He swore that he actually saw a blue sky through the rent. It was nighttime as he gazed through and it was daytime "on the other side." For Gorman, this was a rare glimpse into what might actually be happening on his property. After seeing the blue sky, Gorman began to think that the strange events on the ranch might be explained in terms of different dimensions, alternate realities, and such.

Another night, Tom was again sitting near his favorite tree stump, once again puzzling at the orange structure that hovered in the same area over an abandoned homestead about a mile to the west. He was training his night-vision scope on the middle of the orange mass. This time he couldn't see any sky, but the middle seemed like it had multiple layers, like a three-dimensional onion that moved away from him. And then Tom's sharp eyes picked out a fast-moving black object that was silhouetted perfectly against the bright orange background. The black object seemed to grow bigger, and Tom could tell that it was moving very rapidly in his direction at the center of the orange "window." Within seconds, the vaguely triangular object had gained considerably in size and it appeared bigger as it flew directly toward him out of the "hole" in the sky. The object was moving at such a speed and so quietly that he could make out only the black shape. The object then quickly vanished into the night.

Tom had seen something few humans have reported. He had seen a flying object, black and possibly triangular shaped, flying from a great distance from the "other side" of the orange structure to his property. This incident convinced Tom that his ranch was the site of some kind of dimensional doorway through which a flying object entered and maybe even exited this reality. On another occasion, Gorman saw another object exiting at high speed through the orange rent in the sky. Each time, the flying aircraft moved much too quickly for him to get a good idea of its size or exact shape.

When I asked Gorman for a detailed description of the orange shape, he told me that it looked different depending on the observer's viewing angle. On one occasion, Tom was driving off his property as the orange object appeared. He drove quickly on a narrow road that circled his property, and as he did so, the object became less visible. He reached a point where he simply could not see the object. He then turned around and drove slowly back in the same direction and noticed that the orange object came slowly back into his view. But from the roadway, it looked like a faint orange cloud, almost unnoticeable. As Tom approached his property, he saw the object through his windshield gradually becoming clearer, bigger, and more defined. And finally when he reached the homestead, he understood why other people in the area were not making loud noises about the orange hole in the sky.

The Gorman homestead was the only vantage point from which the orange structure was perfectly visible. It was like a three-dimensional orange tunnel that receded away from them, and the sides of the tunnel were perfectly camouflaged with the sky, so from a side view an observer could see nothing at all. The only perspective that afforded a good view into the interior was directly opposite the mouth of the tunnel. For whatever reason, Tom told me, the "mouth" of the tunnel pointed straight at the Gorman homestead. Motorists passing on the roadway a mere mile away could see only a faint, blurry, orange-colored cloud in the sky. It looked perfectly ordinary.

CHAPTER 9

Vanished

Other than his family, cattle were the love of Tom Gorman's life. He was far ahead of his ranching neighbors in terms of his abilities as a breeder of high-end cattle and even as a cattle rancher. Gorman took it personally if even a single animal disappeared or died. He would be out at 3 A.M. in snowstorms taking care of newborn calves. He was obsessed with good ranch management.

Then in winter of 1994–95, some very weird things began to happen to his cattle. Immediately after a severe snowstorm, Tom would begin rounding up the animals from a far-flung corner of his ranch. That way he could have a head count within twenty-four hours. On this particular night, one of his breeding cows had disappeared in the middle of a snowstorm. He spent almost twenty-four hours on horseback plodding through snow twelve inches deep looking for the registered Angus. It was one of his best and he was determined to find her. He imagined that she might have fallen and broken her leg and was slowly freezing to death. He had searched most of the ranch except for a particularly dense area of trees in the southwest.

Now, rounding a thick copse, Tom sighed with relief. Ahead of him were very fresh tracks in the snow. It would be child's play to track the missing animal now and bring her back. The tracks were plain as daylight and they wound through the dense thickets. As he followed them, Tom's brow puckered. From the spacing of the tracks in the snow and from the bits of snow kicked up, his experienced eye could tell that the cow had begun to run. This was unusual behavior for a cow in the middle of a snowstorm. Usually it would huddle near a thick grove of trees and get whatever shelter it could until the storm abated. What was this animal running from? Gorman knew that predators usually do not hunt during a snowstorm. As he tracked the animal, it became apparent that the cow had been running at full tilt. She had swerved and careened madly through bushes. Her tracks had broken off several sizable twigs and even small branches of the thick vegetation. The weird thing was that the cow's were the only tracks. What was she running from? Tom felt a mixture of puzzlement and dread.

When he rounded a corner and entered a clearing fifty yards in diameter, his breath stopped. The tracks ran wildly out into the middle of the clearing where there was no cover. He felt a cold chill. The tracks just stopped dead in a foot of snow. There was no ambiguity. The animal had just vanished. The last four tracks indicated that the cow was still running at full tilt. The small bits of snow lying around the actual indentations were exactly the same as the previous four imprints. Vainly, he looked left and right, wondering if she could have somehow leaped into a crevice, but there were no crevices. She had just vanished. Gorman quartered the area for half an hour, carefully avoiding the tracks.

What could have lifted a thousand pound cow in full flight off the ground in the middle of a vicious snowstorm? Gorman knew that most helicopters couldn't do it. But something had bodily and dramatically stolen his cow by removing the running animal from the ground. It just didn't make sense.

The other possibility was even more fantastic and unbelievable. Could the cow have simply run through a "doorway"? Tom's mind flashed on the bizarre orange holes in the sky that were becoming increasingly common to him and his family. He hated science fiction. He was a practical man. But he forced himself to ponder the unthinkable. If the cow had run through some kind of "hole," then that doorway had been a one-way entrance.

Tom never saw his animal again.

Over the next few months, another four animals just disappeared. Tom's stress level had climbed dramatically as a result. By April 1995, the very long winter had ended and the heavy rains had begun. For nearly two days the rains lashed the ranch and, as usual, Tom was out there tracking his cattle, along with his son Tad, also on horseback. As the rain hammered down and with only about twenty feet of visibility, Tad was chasing one of the calves that had been born a couple of months prior when he passed a bawling heifer in the canal. She was trying to climb up the muddy embankment, but it was so slippery that she kept sliding back into the water. The animal was plainly distressed and was being vocal about it.

Tad made a mental note to turn back to rescue the heifer after catching the calf. It took him twenty minutes to catch the panicked animal and return it to the mother who was waiting patiently in a sheltered grove. Then Tad returned for the struggling heifer. His heart sank as he saw the animal lying motionless in the canal. In spite of the downpour, the canal was not too deep, so he wondered how on earth the heifer had drowned. He dismounted his horse and jumped the few feet into the canal. What he saw made him retch and to holler for his dad.

The heifer was lying motionless, and her entire rear end had been carved out with what looked like an extremely sharp instrument. Gorman came galloping up at the sound of his son's panicked yells. His face turned ashen when he jumped into the canal with his son. There was no blood in the stream. The cut was flawless. It looked as if a six-inch-diameter, perfectly circular saw with a sucking device had jammed into the heifer's rear end and effortlessly sucked out the entire insides of the animal without any loss of blood. And it had happened right in the middle of a heavy rainstorm.

Tad swore to his dad that he had seen the animal fully alive and healthy a mere twenty minutes before. Both of them climbed grimly out of the canal in silence. Tom took a rope, tied it around the heifer's hocks, and with the rope tied around the saddle of his horse, hauled the dead animal out of the canal. He felt sick to his stomach. He wondered what he would tell Ellen so she wouldn't become hysterical.

The strain of the missing cattle together with the relentless campaign to hide things in her home was beginning to take a toll on his wife. Ellen had begun to feel that something or someone was constantly watching her and waiting for her to leave a room before taking something and hiding it in the microwave. Now this bizarre mutilation had happened with her son and her husband very close by. Tom had heard the phrase "cattle mutilation" before but had dismissed it as the fanciful campfire stories of bored cowboys. He took it very seriously now. He searched the banks of the canal for footprints or tracks, but he knew the heavy rains would have obscured them in a matter of minutes.

Three months later, while riding out early one morning to check his cattle, Tom found yet another mutilated animal. The black Simmental cow was lying near some bushes. He had been worried because he had seen several mysterious lights flying low among his cattle the previous night. The bright yellow headlamps that flew silently and low over the property had become commonplace and created yet more stress and fear for his family. As Tom rode toward the animal, his worst fears were realized. The four-year-old cow's reproductive organs and rear end had been carved out. As he dismounted from his horse, Tom felt a cold anger welling up inside him. An unknown trespasser was violating the sanctity of private property and his family's well-being. He was determined to catch whoever it was. And, if possible, make them pay.

Tom's eye caught a glistening in the sun near the animal's head. One of the ears had been skillfully removed, and right next to the shoulder, a pool of brownish liquid was reflecting the sun on the animal's hide. The pool was about two inches in diameter. He looked closely, and it appeared to be evaporating. He gingerly put his finger into it and a thin gellike substance felt cold. He smelled it and noted a strange chemical odor that he did not recognize. He was determined to try to sample the strange substance so he rode swiftly back to the homestead to get a tightly sealed container. When he returned, it was too late. The material had almost completely evaporated. Gorman swore in frustration.

In early 1996, Tom lost two more animals to mutilation, and this time he noticed that the carcasses were decaying much more slowly than they should have, given the humidity and the temperature. Each time parts of the cattle had been removed with the grim precision of a master surgeon. And each time Tom had seen the puzzling lights flitting around his animals the previous night. He also noticed that the mysterious mutilators seemed to prefer to carry out their handiwork under cover of heavy lightning and thunderstorms. How any technology could reliably function in a heavy thunderstorm was beyond him, but Tom had begun to wonder if he was the victim of some very advanced military skullduggery. Was his ranch the site of some military technology concept testing? Gorman was never sure.

Tom was losing his livelihood, and he felt that finding these perpetrators was now an urgent matter. He was sure that these interlopers were breaking the law. They also had very advanced abilities and used extraordinary stealth to carry out their missions. He was at a complete loss to understand what technology could mutilate animals with such precision.

But Tom's problems with cattle mutilations, although economically devastating, were not unique. They had been widespread throughout the western states, and many eastern states, since the early 1970s.

CHAPTER 10
Mutes

They come in the dark of night. Healthy cattle are silently and skillfully killed and their organs are removed, usually an eye, a tongue, an ear, reproductive organs, and the rear end. We have spoken to many veterinarians who marvel at the precision and skill of the cuts. The cattle surgeons are indeed superb at what they do. From 1975 to 1977, in two Colorado counties alone, there were nearly two hundred reports of mutilated cattle. Governor Richard D. Lamm flew to Pueblo on September 4, 1975, to confer with the executive board of the Cattlemen's Association about the mutilations, which he called "one of the greatest outrages in the history of the western cattle industry." The governor added, "It is no longer possible to blame predators for the mutilations." In the 1970s, in addition to the scores of cases in northeastern Colorado, hundreds, perhaps thousands, of animal mutilation reports were investigated by local law enforcement with cases occurring in fifteen states, from Minnesota and South Dakota and Montana to New Mexico and Texas.

So great was the alarm of ranchers that in April of 1979, ex-astronaut and then New Mexico senator Harrison Schmitt held a one-day hearing on cattle mutilations in Albuquerque. Schmitt began the hearings by saying, "There are few activities more dangerous than an unsolved pattern of crime. There is always the potential for such crimes to escalate in frequency and severity if allowed to go unsolved and unpunished . . . In the last five years—and probably longer—in at least fifteen states animals have been killed and systematically mutilated for no apparent purpose, by persons unknown." In those days, cattle mutilations were taken seriously by politicians, not least because in several states, angry ranchers, convinced that this was an illegal government operation, had begun firing at low-flying National Guard and other military helicopters.

Immediately following the New Mexico conference, a New Mexico state research stipend funded a formal investigation of the cattle mutilation phenomena, and retired FBI bank robber expert Ken Rommel was hired to lead the investigation. Unfortunately, Rommel knew very little about pathology, he knew nothing about the procedures for conducting necropsies, and he entered the topic with his mind made up that this was nothing more than a few ignorant, uneducated ranchers misidentifying perfectly ordinary predator or scavenger attacks. For six months, Rommel sat on the back of a horse and took a few cursory photographs of dead cows, then he filed a three-hundred-page report "showing" that cattle mutilations were actually caused by predators or scavengers. Not a single necropsy was conducted during this period by Rommel's team, nor was even rudimentary pathology carried out. In short, the investigation was undistinguished and lacked any thoroughness. The report was published with much fanfare, "solving" the cattle mutilation mystery, and it provided the perfect rationale for many veterinarians and law enforcement officers to avoid the often disgusting and bizarre subject.

While there are good arguments to be made that a large number of cattle mutilations are actually a covert monitoring operation for infectious diseases—diseases that are deemed necessary by the agriculture and health authorities to keep under wraps for fear of scaring the public—a smaller but undetermined number of cattle mutilations are clearly carried out with a different aim. And if that aim is to strike fear and or terror into the local community, then they have certainly succeeded.

Cattle mutilations are distinguished by the anomalously sudden death of the animals with rarely any signs of a struggle. Normally when an animal dies, even when it is butchered, it will thrash its legs or struggle in the final throes of death agony. Not so with the cattle mutilations. No tracks, human footprints, or tire marks are usually detected on the scene. This lack of evidence is frustrating to the legions of law enforcement professionals who have unsuccessfully investigated cattle mutilations.

Police investigations into cattle mutilations usually fall into one of two categories. Most police feel disgust and intimidation, and are less than motivated to investigate. Others dismiss the mutilation as the work of predators or scavengers. But to a few, animal mutilations are a crime; an affront to a family and its property, and they do an excellent job investigating the cases. Sheriff Tex Graves from Sterling, Colorado; Deputy Wyatt Goring from Cache County, Utah; Captain Keith Wolverton from Great Falls, Montana, Tommy Blann from Texas; officer Ted Oliphant from Fyffe, Alabama; and officer Gabe Valdez from Dulce, New Mexico, are all members of the select group of police officers who did not run away or attempt to ridicule cattle mutilations. And oftentimes they suffered for their bravery and diligence. In spite of their conscientiousness and that of their colleagues, cattle mutilations today remain a law enforcement enigma. Not a single person has been caught or charged in the entire thirty-five year history of the phenomenon.

The paranormal cattle mutilations may have been around a long time. According to researcher Tommy Blann, precise mutilations of livestock have been recorded in England and Scotland as far back as 1810. And Jacques Vallee described an early cattle mutilation that took place in the 1890s when a farmer walked into his backyard and over his field saw a cigar-shaped craft floating about forty feet off the ground. Below was one of his cows dangling by a thick cord. The craft got away, but he found his cow dead and badly cut and burned across the street the next day.

In 1967, a horse called Lady died under suspicious circumstances in the San Luis Valley in Colorado. The San Luis Valley was home to an almost infinite variety of legendary paranormal activity, so it was perhaps fitting that the first widely reported mutilation case would take place there. After this initial incident, the action started in Minnesota and North Dakota and within months had moved to northeastern Colorado.

There are no hard estimates of how frequently cattle mutilations occur, but it is certain that only a minority get reported. The stigma associated with a bizarre, satanic, cultlike phenomenon mixed with a hint of psychological warfare is simply too much for most God-fearing ranching families. It is the exception rather than the rule to go public with these reports. Most just bury the remains of the unfortunate animal and hope that the problem goes away.

Cattle mutilations were not unknown in the Uinta Basin before the arrival of the Gorman family in 1994. Local police had already investigated a couple of dozen cases in the 1970s, many within striking distance of the ranch. Even back in the 1960s, according to local eyewitnesses, there was evidence of cattle mutilations on the very property that the Gormans would move onto. In hindsight, locals remember a series of particularly grisly mutilations in the 1960s and the 1970s on that very same acreage. And the surrounding ranches were not immune.

In fact, the Uinta Basin became such a hotbed of cattle mutilation in the 1970s that well-known mutilation investigator Carl Whiteside from the Colorado Bureau of Investigation even made it a practice to take the trip across the Utah-Colorado border in a helicopter to land near felled animals in farmers' fields. Local ranchers in the Uinta Basin still talk about the Colorado investigative team's unbelievable rudeness and obnoxious behavior. And what was their diagnosis after investigating multiple cases? Predators. But a small number of veterinarians who have had the courage to go to the scene of the mutilations and investigate the cause of death tend to rule out the predator-scavenger theory in favor of something much more sinister. The deliberate killing of the cattle and precise surgical removal of organs has been described on numerous occasions.

So, unknown to Gorman, his ranch was not unique in the long history of the cattle mutilation phenomenon. It was unusual, but not unique, for so many animals to be killed and to disappear from a single ranch in such a short period of time. (When it occurs, it usually involves insurance fraud.) There had been other examples, some well-known and some not so famous, of dozens of livestock being lost to the phantom surgeons from a single ranch over a relatively short period of time. For example, a ranch in northern New Mexico had lost a few dozen animals in the 1990s, and another well-known ranch in northern California had also suffered in excess of thirty animals killed or missing during the 1990s. And a Canadian ranch near Makwa in northwestern Saskatchewan lost more than a dozen animals to mutilations over a two-year period around 1995.

But by and large, the focused decimation of a single herd is unusual. Gorman may have been relatively atypical in having so many *registered* animals mutilated or missing over a fifteen-month period. But the consequences to the Gorman family were both economically and psychologically devastating. They were being harassed on their own property by a ruthless and unseen enemy.

CHAPTER 11

Orbs

Tom and Ellen Gorman stood outside their homestead looking west. It was evening. The summer was coming and the chill of the winter air was no longer apparent. Both were stressed. They watched their cattle grazing several hundred yards away on the lush pasture and, just to the south, three of their horses munched on the grass. Tom could tell there was something wrong: the cattle and the horses were restless.

Tom was the first to see it, as he was quick to notice anything slightly out of place. He stiffened. A blue orb was flying in the tree line next to his horses. He felt Ellen beside him tensing as she too saw it. The intense blue light cast by the object was easily visible as it flew through the trees. They both watched as it emerged from the tree line and slowly flew around the head of one of the horses. The horse noticed it too and impatiently shook his head as if trying to rid a swarm of flies. The darting orb was close enough to illuminate the animal in an eerie bright blue glow. Tom was puzzled by the fact the horse never registered alarm. Normally the blue orbs caused extreme stress in animals.

Suddenly, the blue object darted away from the horses and, with astonishing speed, moved closer to the Gormans. It stopped abruptly in midair about fifteen feet above the ground and hovered silently about twenty feet from them. This was easily the best view they had ever had of the elusive blue orbs. They watched, fascinated, as the object hung in the air, apparently defying the laws of gravity. The exterior of the orb was a clear, hard shell not unlike glass. It was maybe two or three times the size of a baseball. And inside the glasslike exterior, moved a swirling, intensely blue substance. It seemed to Tom like a liquid beginning to boil, a nearly bubbling incandescent blue fluid. He could hear a faint crackling sound from the object like static electricity sometimes makes.

As Tom watched this amazing spectacle, the hair on the back of his neck rose. He could feel a wave of deep, naked fear washing over him. He felt paralyzed with the deepest, most visceral fear he had ever known. It was overwhelming. Wild animals had trapped Tom, he had been close to death, but he had never felt anything like the intensity of the terror he felt now. He knew Ellen was feeling the same because she had begun to hyperventilate. She gasped deeply and her body had begun to shake. Tom felt like he was going to have a seizure.

Suddenly Ellen, who was whimpering with terror, turned on her flashlight. The effect was instantaneous. The blue orb darted abruptly into the branches of the nearby tree as if trying to avoid the flashlight's beam. It maneuvered effortlessly through the branches at high speed. It was obvious to them that the orb was under intelligent control. Then as abruptly as the object had darted into the trees, it suddenly shot out of sight behind their homestead.

Ellen sank to her knees, weeping. Tom also felt weak. His legs could barely hold him. But the overwhelming, paralyzing terror he had felt had vanished. It was like a switch had been abruptly thrown. The aftereffects of that bolt of adrenaline were obvious. Perspiration poured from his body, and his legs and arms began shaking violently. He too sank to his knees and put his arms around his violently trembling wife. She continued weeping. He felt helpless to comfort her and could not manage a reassuring word. As his shaking subsided, he felt only numbness inside. He also felt relief. And puzzlement. How could that orb have provoked such abject terror in both of them?

Tom knew that the fear he had felt was artificial. It had not been a normal response for him. He guessed that this bright blue orb had deliberately manipulated his emotions. How could this be? he wondered. Ellen buried her face is his chest, saying over and over, "We have to leave this place, we have to leave this place." He nodded absently. He knew that his wife was nearing her breaking point. More than a year of relentless psychological warfare by a technology that seemed capable of anticipating their responses even before they reacted had begun taking a deep toll on her and, he knew, his children.

Two hours later, Tom and Ellen were in the living room recovering but exhausted and emotionally spent. Out of the corner of his eye, Tom noticed the signature blue glow outside the window. He stiffened. Ellen gasped in alarm. Both of them watched as it moved slowly past their window, flying lazily. The lights in the living room dimmed as the blue orb flew past, leaving a murky yellow glow inside the house. As the incandescent blue sphere traversed the end of their homestead, the lights inside the house brightened again as if on a dimmer switch. But there were no dimmer switches in their home. They both rushed to the front door in time to see the blue glow floating lazily over the ridge about a hundred yards away. Their bright yard light had also dimmed as the orb moved past and it gradually regained its normal brightness. Neither of them slept much that night. Ellen cried a lot.

The mysterious events were happening thick and fast, against a backdrop of the continuing disappearance of objects both inside and outside the house. Tom eventually found his missing post digger. The only trouble was he found it perched twenty feet up a tree. It would have taken somebody of great strength to lift a seventy-pound post digger up into a tree. The mysteries deepened.

By the end of June 1996, stories and rumors had begun circulating about weird events taking place at a remote ranch in northeastern Utah. Tom groaned. It was only a matter of time before the media got involved and the family's much-prized privacy would be history. One day shortly afterward, as if to confirm their fears, Tom and his son Tad watched a vehicle drive slowly from the entrance gate all the way down to the homestead. As the bouncing vehicle approached, Tom could see a large, blond-haired man at the wheel. Hiding his annoyance, Tom nodded as the stranger dismounted from the vehicle. The guy was broad shouldered and over six foot two. It did not take him long to dispose of the pleasantries. The stranger explained that he had learned about the bizarre events on the property "on the grapevine" and had driven a long distance to visit. Tom interrupted to reiterate that this was private property and that neither he nor his family were interested in developing the land as a tourist attraction. The stranger was insistent, even pleading. All he wanted to do, he explained, was to go onto the property and meditate. Tom could see his son grinning to himself, and eventually, half in amusement at the bizarre request, Tom relented.

The three of them piled into Tom's diesel truck and headed down into the ranch. After about a mile, the stranger announced that he would like to meditate here, near a small pasture surrounded by trees. The stranger walked into the middle of the open ground, about a hundred yards from the tree line. Tom walked with him a short distance and then stood watching. He glanced back at his still grinning son who had elected to stay by the truck. Tom was about thirty yards from the stranger, who had closed his eyes and, in a faintly religious gesture, had spread both his arms out. Tom was amused.

Silence reigned and the late afternoon sun cast a beautiful light on the scene, this tall blond man standing silently in the middle of the pasture with his eyes closed and his arms raised, much like the pose struck by saints and angels in religious paintings. In the distance Tom heard the sudden chime of a cowbell. He was puzzled. None of his animals had cowbells. The sound seemed to be coming from deep within the trees. There it was again, nearer this time. The stranger seemed not to have heard it. Tad made a gesture of puzzlement. Tom looked at the trees and thought he could see a faint blur. Something was moving very quickly between the trees. Tom could not make out the shape, but he knew it was big. Was that the source of the cowbell sound? He watched carefully as the shape moved like a fast blur from tree to tree. It was almost as if it was circling. Tom suddenly felt uneasy.

Without warning, something broke from the tree line and moved swiftly toward the meditating man. Tom blinked. He still couldn't see what it was even though it was broad daylight. It was blurred as if it was hidden in the middle of heat distortion, and it was covering ground at enormous speed. Gorman realized that this chimera was making a beeline for the blissful meditator, who was completely unaware of what was rapidly bearing down on him. Tom was about to yell a warning, but it was too late. The shimmering wraithlike huge "thing" had stopped just inches from the meditator as it let out a deep-throated animal roar that echoed around the ranch. The roar sounded half like a bear, half like a lion. Tom froze.

The stranger leaped back about ten feet and fell down. He began screaming. As fast as it had approached, the shimmering, almost invisible "creature" departed for the tree line at top speed. Tom's sharp eyes could make out only a blur of dancing, flickering, wavy lines, like pixilated blocks. Within seconds, the creature had vanished into the trees.

The visitor was on the ground, still screaming hysterically, and Tom hurried over to make sure he had not been injured. Suddenly, the stranger jumped up and threw his arms around Tom, weeping like a baby. He was obviously out of his mind with fear. Tom struggled to extricate himself, but the guy was big and he was possessed of a strength borne of blind panic. He simply would not let go. After a few minutes, Tom said quietly, "If you do not let go, I am going to hit you."

"I will let go, if you promise to get me to my vehicle," the stranger babbled. His ruddy face had turned chalk white and it was obvious he felt the fear of God. Slowly, with the man still hanging on to him, Tom made his way to the truck. His son started the engine and Tom, with his cargo of blubbering humanity, climbed into the back seat. The stranger swore that this property was cursed and that he would never set foot on it again. Of that, Tom was thankful.

They watched as the stranger drove erratically toward the gate. He was driving dangerously fast on the rutted track and Tom hoped he would slow down once he got on the country road. Tad was still shaken. That roar had penetrated to the very core of their being. It was like being shot with a bullet.

Some time later, as Tom and Tad were watching the movie *Predator*, in which Arnold Schwarzenegger and Jesse Ventura battle an alien life-form in a jungle in Central America, they let out a loud yell when they first saw the shimmering creature. "That's what we saw," they yelled in unison to the astonished family. The predator in the movie seemed to exactly encapsulate the degree of camouflage of what they had seen. Tom calculated that the thing they saw was moving at between fifty and sixty miles per hour when it broke cover from the trees.

Again Tom wondered if his ranch had become a testing

ground for advanced military equipment. Now suddenly the list had expanded beyond high-tech aircraft and surgical derring-do to advanced camouflage technology. But why would some kind of super tech advanced vehicle emit a roar? Was it possible that a creature could have advanced camouflage capabilities? Like a chameleon taken to the next level? Tom refused to dwell too deeply on what that creature might do if it decided to harm his family.

He already knew what the creature was capable of.

On one April evening in 1996, Tom sat outside, looking west and trying to relax. The beauty of this property was undeniable, yet the family was feeling overwhelmed by the stress caused by the strange phenomena. His three dogs sat contentedly beside him. Tom was grateful for such loyal dogs. All three were heelers. They were gentle with the family but absolutely ruthless with strangers. They were also aggressive cattle dogs. The cattle knew when to obey these animals, and he could rely on them to defend his valuable, artificially inseminated breeders against the numerous coyotes, raccoons, and wild dogs that moved through his property. All in all he was feeling almost content.

It had become a regular occurrence now to see a large orange something hovering slightly above the cottonwood trees about a mile west of where he sat. It did not particularly concern him, as he had seen these large orange things dozens of times before. Gorman had spent hours looking at them over time through the rifle scope that he carried with him to enhance his already superb eyesight. Very occasionally, Gorman had seen objects flying out of these orange things, as if they were windows into another dimension. He had little understanding of the physics involved. All he knew was that he wished they would go away.

Then he spotted an object in the distance at the far end of the pasture. The sight gave him a chill. A small flash of intense blue. Tom straightened in his chair, all pretense at relaxation gone. The dogs also took notice and began their low-pitched growls.

He saw it again, and it was only three hundred yards away, moving swiftly along the bottom of his pasture in a north-south direction. It was less than ten feet off the ground. When it got to the southern end of his pasture, it abruptly turned and began flying in Tom's direction. He tensed. He could see it much more clearly now: a perfectly round, intense blue orb, bigger than a baseball and capable of very sophisticated, intelligent maneuvers. He had seen them so many times on his property before. And they usually signaled trouble. His dogs were barking. The object was now less than a hundred yards away and it had changed direction again. It was moving north, parallel to Tom's position.

Without really thinking, he set his dogs loose. Usually he kept them beside him when these things were flying around, but tonight he lost his patience. His three dogs took off at top speed in the direction of the blue orb. It didn't seem to react to their presence until the animals got much nearer. Then it dipped down and descended until it was only a few feet off the ground. The three dogs began leaping at the object. They were snarling with jaws snapping. Each time the animals leaped at the orb, it skillfully moved out of the way, the jaws missing sometimes only by inches.

This strange ritual began to play itself out. It was apparent that the dogs were becoming incensed with this strange object that danced out of the way at the last moment but then dipped down again so that they could lunge at it again. The intense blue orb seemed to be deliberately teasing the enraged dogs.

Tom was getting increasingly uneasy as the game of catch moved in the direction of a thick copse of trees a couple hundred yards to his south. He sensed that the orbs seemed to be steering the dogs toward the cover. A couple of minutes later the orb dipped to the ground and with almost languorous speed flew slowly among the trees. The snarling, eager dogs gave chase. Suddenly, Gorman heard sounds that chilled him to the bone: the unmistakable fear-filled yelps of dogs in mortal agony. Then an eerie silence. Nothing moved. Tom waited for his animals to return. After a couple of hours, he went into the homestead with a heavy heart. He decided not to look for them until morning.

His worst fears were realized when he went down the following morning to inspect the copse of trees. A smell of burned flesh greeted his nostrils as he dipped his head beneath the low branches. Ten yards inside was a small clearing. Tears filled Tom's eyes. Three large circles of brown, dried-out grass were in the middle of the clearing. At the center of each circle of shriveled vegetation was a blackish greasy mess. The stink of his incinerated dogs was awful. Tom rushed out of the copse, his mouth dry and his stomach heaving.

Within hours, Tom had gathered his family and finally agreed, as they had insisted, to sell the ranch. The ruthless killing of his favorite companions was the last straw for Tom, who after fighting an unseen, unknown enemy, reluctantly surrendered. By July 1996, the family had had enough. None of them had slept well for months. Now the Gormans finally understood the weird arrangement of dead bolts they had found throughout the house a mere twenty months before.

The story of the Gorman ranch had hit the newspapers and it ricocheted around the country. It wasn't long before it caught the attention of one of the most powerful businessmen in North America and his organization, the National Institute for Discovery Science.

PART II

THE INVESTIGATION BEGINS

CHAPTER 12
Science

The National Institute for Discovery Science (NIDS) was an ambitious undertaking. Never before had a scientific organization been created and, more important, funded in order to bring scientific rigor to what was essentially paranormal research. When I saw the NIDS ad looking for interested scientists in the prestigious journal *Science*, I leaped at the chance to do revolutionary science in an area where few scientists had previously ventured. In answering the recruitment ad, the words of one of my postdoctoral mentors came to mind: "If you want to catch a big fish, the best way is to fish in waters that are unpopulated by other fishermen." Within a few months of joining this brand-new organization in the summer of 1996, I began to see NIDS as a way to roll back the frontiers of discovery science while working with some of the most brilliant people I had ever met.

In addition to the two PhD scientists working with me, NIDS had a world-class, multidisciplinary advisory board that had been carefully hand-picked from an array of disciplines in mainstream science. NIDS wanted to bring as broad a range of expertise and technology as possible to bear on UFOs and other problems that mainstream science had thus far ignored. The study of so-called paranormal events had always been starved of funding, and NIDS, which was started by Las Vegas real estate tycoon Bob Bigelow, was designed to remedy that situation.

Shortly after NIDS staff was hired, we heard about what was happening on the Gorman ranch. The opportunity it presented seemed ideal. The idea of a field station, or "laboratory in the wild," that would bring all NIDS resources to bear to study it gained traction and quickly came to fruition. Within a few short weeks, NIDS bought the Gorman ranch in Utah.

The Gorman family was finally free to leave the property on which they had been held hostage for more than eighteen months. They purchased a small ranch twenty-five miles away, and Ellen, Tom, and the children quickly made it their home. At the time, they told me that they wouldn't care if they never set foot on their old ranch again. There was fear mixed with revulsion in their reaction. Three of the Gormans were supremely happy to move on and to bring this part of their lives to a close.

But Tom Gorman was not quite ready to do so. He is a proud man and was outraged at being thrown off his land by something he didn't understand. In the space of eighteen to twenty months, someone or something had killed or stolen fourteen registered cattle out of a herd of eighty animals, an attrition rate approaching 20 percent. Each of the animals was worth a couple of thousand dollars. Economically, the family was devastated. Even with such high-end cattle, the profit margin for ranching was slim. But in addition to the financial losses, Gorman's family had been reduced to barely surviving automatons because of stress and sleep deprivation. Their kids' grades had suffered drastically in school and they were the butt of jokes in the community once the media took an active interest in their story.

Within a couple of weeks of the transfer of ownership, Tom decided to take a job as the NIDS ranch manager to help us come to grips with who or what had run him off his land. NIDS purchased a few dozen cows and Tom kept several more on the property to be used as bait. With remarkable self-discipline, Gorman arrived at what had been his property every day. Once he had been owner, now he was an employee. He took care of the animals, supervised the irrigation, mended the broken fences, and joined the scientific teams on watches that sometimes lasted all night. On other nights, Tom returned to his family. The pace was grueling for Tom. While his family rested, his stubborn refusal to allow the phenomenon to get the best of him drove him onward. He was determined to find out more about what had terrorized his family.

I will never forget my first sight of the ranch. The team had flown up in early September 1996 to reconnoiter the area before moving the observation trailer/laboratory onto the property. It was a clear, sunny day as we drove the narrow half-mile track that ends up at the homestead where the Gormans had suffered so many sleepless nights. On our right, the red sandstone ridges jutted into the sky. I was struck by the wildness and pastoral beauty of these Utah badlands.

When I first set foot on the ranch I had the unmistakable feeling that something was not quite right. Things were not what they seemed. Everything looked beautiful as Tom took us on the tour on that first day. We briefly met Ellen and the kids. They looked exhausted, with white faces and dark-rimmed red eyes. Only later would I learn from Tom the extent of the trauma that his family had suffered.

The ranch looked spectacularly gorgeous that day. The trees had not yet shed their blooms and the cattle grazed peacefully in the fields. At the same time, in marked contrast to the serene environment, I could not shake an eerie feeling. Was I being watched? As a scientist I immediately put this feeling down to the hype and the expectation about the place, but as a human being I knew the feeling went deeper than that. Needless to say, I did not mention it until much later.

Tom then took us on a tour. We began walking west, and the first stop brought us close to his fence line. He showed us the carcasses of two of his neighbors' cows that had gone missing a few days previously and had just been found in unlikely positions beneath the barbed-wire fence that separated the two properties. Even from fifty yards away, the air was heavy with the stench of decaying flesh. The sound of thousands of buzzing flies as they landed in the open mouths, sightless eyes, and every available orifice and exposed soft tissues of the carcasses was almost deafening.

The animals lay within thirty yards of each other with their heads under the wiring of the fence. The NIDS veterinarian thought this was an unusual position for two animals to be in at death. There were few signs of struggle near the stinking animals. We looked carefully around the vicinity, and there were no marks.

As we approached the animals, I noticed movement and I was momentarily puzzled. At first it looked like large white tarps or white plastic bags moving in the breeze near the cows' bellies. As we got closer, I realized that I was seeing a flowing river of millions of white, glistening maggots pouring from the abdomens of both carcasses. Watching them was like seeing a single giant organism writhe and swivel in and out of the abdominal cavities. Both cows had been dead about forty-eight hours, judging by the number of maggots. Forensic scientists agree that generally it takes only a couple of minutes after death for the first blowfly to land and lay eggs on the carcass of a dead animal. Within less than a day, the first maggots appear. Meanwhile, more blowflies quickly find the dead animal and rapidly repeat the egg-laying process. At the same time, the gas from the fermenting grass inside the rumen of the cow begins to expand and bloat the belly. In the high heat and high humidity of the Utah summer, this process happens very quickly. It is never a good idea to prod or kick the swollen, bloated abdomen of a dead cow. You might cause an extremely unpleasant and smelly explosion.

Whether these two animals had died of natural or other causes was open to question since no obvious tissues were missing and the animals were far too decomposed to gain any biologically useful information from a necropsy. I was thankful for my previous experience of being around dead cows, during my experimental work on cattle in a lab in France and on another stint at a large animal facility in Ottawa, Canada. But the physicist who accompanied us on the tour was not so fortunate. A career of writing software for NASA deep-space missions had not prepared him for an up close and personal encounter with a thousand pounds of putrid, maggot-infested flesh. Within seconds, his face was turning a deep shade of gray and I noticed his vain efforts to disguise the retching in the back of his throat.

Tom took us to a small clearing about a hundred yards west of the two dead cows. Even at this distance, the acrid smell still wafted on the slight breeze. Behind a few trees, we came upon three circles of very dried grass. All around the twenty-foot-diameter circles of the desiccated grass, tall healthy green grass flourished in three-foot-high swathes. This is where Tom's three dogs had been incinerated, he explained, still plainly upset by the memory.

As we walked on, Tom pointed out the remains of two more cows that he said had been mutilated several months previously. A shell of weathered, tanned hide surrounded both carcasses. Each animal's bones were plainly visible, but their rear ends were missing. Since it was months after the fact, no useful information about the manner of their death could be ascertained. Tom explained that it had taken far longer than usual for the two carcasses to decompose. Usually, with the kind of temperatures and humidity in northeastern Utah, the carcasses would be bones within a few weeks. But it had taken almost a year to reduce these animals to this condition. Tom asked the veterinarian with us what could have slowed the rate of decomposition. The veterinarian shrugged. Perhaps a dramatic decrease in temperature, or a chemical had killed off the putrefying bacteria. Neither seems very likely. It was easier to believe that Tom had got his facts wrong. This was not to be the last time that NIDS would underestimate what Tom told us. We walked on.

Five hundred yards west, he showed us two circular, deep holes in the ground where soil had been removed. Tom had found several of these deep indentations, usually following a night of flying lights. The holes were about a foot deep and several feet in diameter. They sloped downward from the edges. I guessed that two or three hundred pounds of soil would have had to be removed from the ground in order to make those holes. I asked Tom if he had found soil nearby. He shook his head. He explained that the first time he had seen the holes, they had perfectly straight sides, as if a giant cookie cutter had dug in and removed the soil and grass. Over time, he explained, weathering and rain had blurred the precise cuts of the holes. Now, with grass beginning to grow, the holes looked obvious, but less than spectacular.

The NIDS plan was a simple one. Before anything could be accomplished, a validation process had to occur. A science team needed to personally experience the anomalous events. In other words, they had to see anomalous phenomena with their own eyes. That was the first order of business. We estimated that this phase would take a few weeks or months. Little did we know.

CHAPTER 13

Approach

A dramatic change occurred at the Gorman ranch in the fall of 1996. The hunter became the hunted. The phenomenon was suddenly confronted not with a family but with a scientific team whose primary motivation was not fear but curiosity. This was a significant turn of events for something—whatever it was—that had, for all intents and purposes, been in charge for several years. Now it was no longer in charge.

But how exactly would the scientific team carry out this highly unusual research? There were two sides to the issue. One camp argued for full instrumentation and to cover every square yard of the property with automated sensors that would constantly feed back data. The other camp argued that less is more, that too much equipment, too much activity, might even be detrimental. That was Tom Gorman's point of view.

Tom believed that too much activity and technology was guaranteed to drive the phenomenon into hiding. He thought the NIDS group should set up a command post in nearby Roosevelt or Vernal and silently and surreptitiously creep onto the property at night while disturbing as few geographical landmarks as possible. "This phenomenon needs to be hunted like a wild animal," he told me on numerous occasions. "Maybe even a very smart big game animal."

Tom also recommended a stealth method for obtaining video footage of the mysterious flying activity that haunted the ranch. He used to sneak out of his house after dark, armed with an old manual video recorder that contained as few electronics as possible. Slowly and painstakingly, often on his stomach, he would make his way down to a vantage point where the weird floating lights were often seen. That spot was about two-thirds of a mile west of his house. Like an expert hunter, he would wriggle along the ground while taking care not to break any branches or twigs, and he would remain in the same place for twenty minutes if he thought he had made the slightest noise. Sometimes he would take hours to arrive at a good location where he could observe and listen.

Tom told me he had lain almost frozen in ditches for hours while he waited to catch a few minutes of activity on videotape. Though only marginally successful, he did see numerous quietly floating lights of all shapes and sizes during these nightly forays. Tom assured me his methods were successful and had produced results, and this proved to him that whatever was terrorizing him was not necessarily omnipotent. Or maybe it was just playing with him. Maybe it could see him perfectly and was just reeling him in like a fish. Tom said he never really knew.

Whether to deploy more or less equipment and personnel was a hotly discussed issue within NIDS. There was little precedent on which to base our decision. Some members of the team researched the two best-known areas where scientific equipment had been deployed to study strange phenomena: Project Hessdalen in Norway and Gulf Breeze in Florida.

The Hessdalen Valley lies in central Norway and has been the site of numerous unexplained lights for decades. In 1984, engineer Erling Strand and others carried out a thirty-six-day instrumented coverage of the area. A magnetometer, a radio spectrum analyzer, a seismograph, cameras (some with dispersion gratings), a Geiger counter, an infrared viewer, and a laser were deployed in the area for about seven weeks as a "pilot" experiment. This short experiment established that the Hessdalen phenomenon was measurable.

The major finding was a correlation between the appearance of luminous phenomena and magnetic perturbations. Attempts at obtaining line spectra through diffraction grating analysis were unsuccessful, although fifty-three visually witnessed occurrences of unexplained lights happened and a few photographs were taken. This initial success was then translated into an expanded and automated set of measurement protocols.

In the 1990s, under the guidance of Erling Strand and Bjorn Gitle Hauge, who by then were assistant professors at Østfold College in Sarpsborg, a real-time automated observatory, dubbed the Automated Measurement Station (AMS), was designed and built. Beginning in 1998, the AMS was deployed in the Hessdalen Valley. The AMS was equipped with automatic wide-angle and zoom video cameras able to monitor the phenomenon in real time, as well as a radar transponder and a magnetometer. During five years of operation, the AMS recorded an extremely valuable statistical variation of the lights, and it indicated that the lights were not from any mundane man-made source.

Early on, there were indications of correlations among sunspot activity, geomagnetic storms, and the appearance of the lights, although subsequent, more intensive analysis ruled out that correlation. The phenomenon was recorded more often in winter and between the hours of 10 P.M. and 1 A.M. "However," the study author noted, "these statistics . . . did not lead to an understanding of the origin or nature of the phenomenon."

Project EMBLA (Electro Magnetic Behavior of Luminous

Anomalies) followed. EMBLA was a joint mission of Italian radio astronomy groups, the Italian National Research Council, and the Norwegian team. The aim was to expand the instrumentation, and three missions were deployed to the Hessdalen Valley. Thus, in August 2000, the Norwegian AMS station in Hessdalen was equipped with an additional set of automated instruments from the Italian group. These included a VLF-ELF (very low frequency–extremely low frequency) correlation receiver and spectrometer; a VLF Inspire receiver; two spectrometers; (Sentinel 1 and Sentinel 2), both centered at 1420 MHz (this frequency is used in astrophysical and SETI studies); and a wide-band antenna connected to a spectrum analyzer. All instruments in the Italian package were computer controlled. Data was recorded continuously and automatically and stored on CD-ROMs.

"The global picture of the phenomenon obtained so far," the Norwegian-Italian collaborative project concluded, "shows that the phenomenon's radiant power varies, reaching values up to 19kW. These changes are caused by the sudden surface variations of the illuminated area owing to the appearance of clusters of light balls that behave in a thermally self-regulated way. Apparent characteristics consistent with a solid are strongly suspected from the study of distributions of radiant power. Other anomalous characteristics include the capability to eject smaller light balls, some unidentified frequency shift in VLF range, and possible deposition of metallic particles."

But the bottom line of Project EMBLA was the same: "A self-consistent definitive theory of the phenomenon's nature and origin in all its aspects cannot be constructed yet quantitatively, but some of the observations can be explained by an electro-chemical model for the ball lightning phenomenon." A rough, albeit less elegantly phrased, translation of the above statement could be: "We don't know what the hell is going on."

As the chief field research scientist at NIDS (members of the esteemed science advisory board were usually not part of the field research teams, although they participated on occasion), I and others spent hours discussing the Hessdalen project with Erling Strand and his group. What emerged is glaringly absent from the scientific discussion of the Hessdalen instrumentation packages and research program. When I asked him, "What do the local residents of the Hessdalen Valley experience?" Erling answered that there were multiple reports of bizarre encounters with UFOs, and reports of abductions and other odd events had been commonplace in the valley for decades. The engineers in the AMS project had little motivation or interest in pursuing this angle. They focused, quite rightly, on perfecting the ability to measure these phenomena, not delving into the phenomenology and sociology of the reports. But I wonder if the locals might not have provided the group with fresh insight into what was going on in the valley. There was obviously an unbridgeable gulf separating the engineers and their instrumentation from the soft squishy stories with no physical evidence of abductions, "huge triangular" craft, UFO lore, strange creatures, etc. In August 2004, Project Hessdalen was suspended due to financial difficulties.

Gulf Breeze is the other hot spot that received scientific attention for a brief period of time. An unprecedented series of sightings took place in this town on the Florida panhandle between November 1990 and July 1992. "During that time," reported physicist Bruce Maccabee, "the Gulf Breeze Research Team (GBRT) logged about 170 sightings, most of which involved multiple witnesses and most of which included still photography with telephoto lenses and/or recording by video cameras. In several cases a light was observed simultaneously by two separated groups of people thereby allowing for triangulation. In one case infrared sensitive film detected a change in the output radiation from a light and in another case a diffraction grating was used to obtain a spectrum of a Bubba UFO [the unknown flying thing got its name when the Skywatchers in Gulf Breeze kept shouting,

"Look over there, Bubba!"] and also the spectrum of a red road flare. The spectra were found to be different."

The sightings at Gulf Breeze continued for a few months and provoked enough interest that a UFO detection van was assembled in Canada and the United States by a private engineering firm. The van rolled into Gulf Breeze, and a few nights later, the activity stopped. Only a couple of anomalous magnetic field measurements were made before the UFO activity terminated. The attempt to instrument the area was eventually aborted because of a lack of activity.

The full impact of these studies would elude us for several years.

In early September 1996, the NIDS team, which was then composed of a physicist, a veterinarian—both of whom do not want their names revealed—and myself, moved into an observation trailer that had been rapidly deployed on the property. We had begun to execute our plan—against Tom's better judgment. Our aim was to gather data in the electromagnetic and magnetic regions, as well as a visible UV spectrum of any UFO lights. To accomplish this, over the course of the next few months we assembled a light-gathering device with a Fresnel lens. This large-diameter lens was designed to focus light onto an optic fiber, which in turn fed directly into a portable handheld spectrometer purchased from Ocean Optics. The spectrometer was literally palm-sized and was linked to a laptop computer. This immensely portable station was ideal for investigating different areas of the ranch and was used to gather real-time spectra in UV and visible ranges. The spectra were then stored on the laptop to be analyzed as needed. In addition to this, the team had a very portable assortment of night-vision binoculars, video cameras (with night-vision attachments), radio frequency analyzers, microwave detectors, etc. We also hired a couple of additional investigators.

In the early phase of the NIDS project, two teams were deployed on the property every night. The teams communicated with Motorola walkie-talkies. Each team had at least one scientist and one or two investigators. The teams' mandate was to capture evidence of any unusual events on videotape and cameras. This way the data, if any, could be impartially scrutinized and critiqued by the fifteen-member Science Advisory Board. During this initial start-up phase, the advisory board regularly flew into Las Vegas for intensive two-day briefings by the scientific field staff.

On September 16, 1996, at about 1:30 A.M., the team had moved into the observation trailer to take a break, when out of the window somebody spotted a light over the cottonwood trees at the west end of the ranch. I was there with Tom and two scientists. The light was so bright that we at first thought it might be a flare. It hovered for about ten minutes above the distant tree line, moved down out of sight, and then back up. We all agreed that the behavior and the appearance of the light were unlike any aircraft, helicopter, flare, star, or planet. In short, the object was unidentified. Cameras, night-vision devices, and digital video images did little to resolve the distant light. We took several photos that showed nothing but a tiny distant light.

Nevertheless, the appearance of this distant object was cause for excitement among the science crew. Though it was a good distance away, it was definitely unidentified. And it was a tiny first step on the road to confirming what Tom had been seeing. But we agreed that the sighting, although unexplained, was also profoundly mundane. It did not really qualify as an independent validation of Tom's observations as defined by the NIDS plan. The sighting did serve as an excellent morale booster, however, and the night watches were redoubled.

We felt we were ready for anything.

CHAPTER 14
Cat and Mouse

Week after week during October and November of 1996, the team flew from Las Vegas to Utah to conduct night watches and to begin the process of interviewing locals in the area. The Ute Indians were very cooperative and gracious. It turned out that several tribe members had multiple experiences with weird flying objects in the previous decades. In other words, what had happened on the ranch was not isolated. It was a part of an overall pattern.

NIDS personnel also interviewed neighbors who were not Native Americans and found them polite but reserved. After several weeks of visits, they began to talk. They had experienced the same thing as their Ute neighbors had. The ranches bordering on what was now the NIDS property had had their fair share of weird activity, so whatever was going on at the ranch was also common in the area. In the early months of this project, NIDS assembled a library of dozens of tape-recorded interviews from local basin residents about strange events that had occurred over

the years near the ranch, including cattle mutilations and sightings of multiple colored balls of light and a large triangular-
shaped object.

Few locals were willing to talk about these experiences on
the record, however. One neighbor, whom we will call Mr. Gonsalez, explained how he lost many cattle over the years. Back in
1995, he remembered finding a recumbent cow lying out in a
field where she shouldn't have been. The rest of the animals
were in a field nearby and there were no broken fence lines. The
old rancher described going out to his eight-year-old cow and
finding that she had two broken legs. Alarmed, he ran back inside to get a blanket to cover the shivering animal. The cow was
obviously suffering and in shock. He suspected he might have to
put her down.

Gonsalez was astonished when he returned with the blanket
five minutes later to discover that the animal was gone. He
looked everywhere but couldn't find her. The field was an open
pasture with no rocks or trees behind which an animal could
hide. Yet in the space of a few minutes, in daylight, a cow with
two broken legs had vanished. An hour later he looked out his
window. It was now afternoon. He told me he nearly fainted
when he saw the cow lying in the same field but about fifty
yards from her original position. He ran out to the suffering animal and examined her closely. This time all four legs were broken. He ran inside to get his gun and quickly put the poor animal
out of its misery. After thinking long and hard about this bizarre
incident, Gonsalez concluded that the animal must have twice
been lifted into some aircraft and twice been dropped into the
field. Each time two of her legs had been broken. This was the
only explanation that seemed to fit the facts. We didn't argue
with him.

In our many hours of conversation, Gonsalez and his family
related dozens of odd incidents to us. These incidents, though
they differed from the Gormans' own, convinced us that the

ranch was by no means a unique piece of real estate in the area. The family told us about strange Mexican-hat-shaped flying objects that flew and hovered over the ridge a mere hundred yards from their home. Mrs. Gonsalez told us that she was returning one night from the small town of Fort Duchesne, when she saw a fast-moving silvery object that rapidly descended in the direction of the red-rock ridge near her home. The object accelerated as it neared the ground, and she waited fearfully for the explosion that would surely engulf her. Instead, the silvery aircraft smoothly flew into the ridge as if it didn't exist. She told her family about it when she returned home.

On the evening of November 10, 1996, my phone rang. It was Tom Gorman. The NIDS team was due to return to Utah the next day for yet another stint of stargazing by night and patrolling the ranch and interviewing locals by day. In his usual gruff but very precise way, Tom described what had just happened to him. He was knocking off work when he saw three yellow-colored headlights flying in close formation only a few feet off the ground near the southwestern border of the property. Cattle were rarely grazed there. The lights looked like those he had seen countless times before.

Early the next morning, the NIDS scientific team was in the air, on the way to Utah to investigate the phenomena. Tom met us, and immediately we trekked through the dense vegetation to the spot where he was sure the object had been. We had with us an array of small portable detectors. The team, as planned, split up and began quartering the several-hundred-square-yard desert looking for any tracks or anything out of the ordinary. We spent several hours doing it but saw nothing. At the same time we scanned the fine sandy soil in scores of places for any traces of nuclear radiation, for any magnetic field signatures. We found nothing. Wearily, we returned to the observation lab. That night we deployed in one area for several hours, then gave up and deployed on the eastern end of the property for several more hours. We saw nothing.

Three nights later, on November 13, at 1:30 A.M., I was deployed with one member of the scientific team close to the observation trailer. We had been watching the area for several hours. Another team was deployed down at the western end of the ranch. We communicated by walkie-talkies only when absolutely necessary. We were both looking at the night sky when suddenly, out of nowhere, a silent, bright yellow light came speeding out of the night from over the lip of the ridge. It was moving as fast as a high-speed jet aircraft but made no sound.

As we watched in shock, the object did a perfect 360-degree circle right over us, again in complete silence, and then zoomed rapidly north. As it headed back over the ridge, I managed to quickly take a couple of photographs. In seconds the flying mystery was gone. The photos, later developed, only showed a dimly visible, blurry light. The fast, silent object had caught us unawares and departed before we could spring into action. Nevertheless, the sighting by two members of the scientific team represented a validation. But again, the event was much too transient for any meaningful interpretation. At night it is extremely difficult to accurately estimate either altitude or distance, and this object proved no exception.

The snows then came to Utah. The NIDS team stayed on regular night watches until the end of November and then, as the temperature dropped below zero, we returned to Nevada and remained on call. December 1996 passed without event and northeastern Utah entered into a deep freeze. We had telephone briefings and conversations with Tom at least weekly. Because of the extreme temperatures, he had pulled all cattle except a few calves in the corral off the ranch and was feeding them at an indoor facility. Tom told us to keep away until the temperature started climbing up again in March.

But on January 21, 1997, something very strange happened.

Tom called us the next day and reported injuries to the ear and eyes of three calves in the corral near the observation trailer. The injuries occurred during an intense snowstorm when the temperature was about 30 below, too cold even for predators, Tom assured us. We told him to call the local veterinarian and take some photos.

The photos showed three small pathetic calves huddled in the thirty-below weather. One had an ear cut up as if with pinking shears. The other two had small round holes punctured in their eyelids. Because of extreme weather conditions, and the vets' reluctance to venture out at thirty below, it took Tom about twenty-four hours to get qualified medical examiners to the ranch.

When they finally arrived, one veterinarian said the wounds were strange and unlike anything he had seen before. This vet agreed with Tom—predators would not conduct attacks on livestock in a corral at thirty below. The second, more senior, veterinarian overruled him and insisted that it was a simple coyote or cat attack. The idea of coyotes or cats attacking calves in a corral next to a homestead in the middle of a snowstorm at thirty below struck Tom as comical.

A heated discussion took place in private between the two professionals. When the two men returned, the older vet simply said that a predator attack had occurred and to keep an eye out for further attacks. The younger kept silent and tight-lipped. "Welcome to the ambiguities in the veterinarian profession as they are forced to deal with anomalous injuries and deaths to animals," Tom told us. "The majority of vets want nothing whatever to do with cattle mutilations, unexplained injuries to animals, or anything that might provoke gossip in the local community that might, in turn, adversely affect their business." But as we gained more experience in dealing with vets, it became obvious that a small minority were willing to follow the data, even at the risk of ridicule from their peers.

On February 21, Tom brought his cattle back onto the ranch. He had told us that the act of bringing his cattle onto the property for the first time, back a couple of years ago, seemed to trigger an escalation of anomalous activity. So the NIDS team was expectant, waiting for the penny to drop. We did not have long to wait.

On March 10, I was working on a project in the corporate office in downtown Las Vegas when Tom phoned. I could tell by the edge in his voice that he was very disturbed. "They got a newborn calf," he said hoarsely. "We were close by and we didn't see or hear a damn thing." Tom was one of the most collected, together people I had ever met, and here he was close to babbling. As I tried to calm him down, the story came out. One of his valuable Black Angus calves had just been dismembered in broad daylight.

I had a gut feeling that this might be one of those cases we were waiting for. I immediately made the necessary arrangements and less than ninety minutes later a private jet was waiting on the tarmac at McCarran Airport to whisk us to northern Utah. No other group of scientists investigating the paranormal had a private jet at their disposal. Most of them couldn't even afford to rent a car for the weekend.

CHAPTER 15

The Killing

Hunched in the plane at thirty-five thousand feet, I looked out the window at the rolling clouds below and wondered what we were getting into. The aircraft hummed as we sped toward Vernal. There were three of us—the veterinarian, the physicist, and myself. We had all the equipment we needed, including all the knives and scalpels necessary to further dismember the animal during a necropsy.

Just five hours after Tom's frantic phone call, we were standing over the animal. The late afternoon breeze blew gently, but an early spring sun was still warm on my face. We were looking at a scene of horror. I felt a churning in my stomach as I looked at the creature. This was something truly bizarre. My immediate impression was that an enormous force had ripped the animal apart. One of the leg bones was lying ten feet away, having been yanked free of the knee joint. Even with a young calf, the brute force necessary to rip a femur off a knee joint and snap a tendon suggested something very powerful.

Yet there was a fastidious delicacy to the way the mutilated calf had been carefully laid out on the grass with all four legs spread neatly away from the body. I had a momentary image of a huge amorphous creature carefully laying a limp rag doll on the grass and gently placing each of the lifeless limbs away from the torso, arranged with the finesse and attention to detail of a Japanese tea ceremony. I shuddered and banished the image from my mind. The combination of overwhelming force in ripping the calf apart and dainty precision in laying the body on the grass seemed all wrong. It disturbed me.

There was no smell. The inside of the animal looked pink and tender, very healthy and very clean, almost unnaturally clean. All of the internal organs were gone and the broken ribs jutted forlornly toward the sky. The head lay sideways, its lifeless eyes staring toward the western sun now low in the sky. We estimated that this was an eighty-four-pound calf, at least forty pounds of which were gone, if you counted its three liters of blood.

And this was the most chilling part of the scene—the complete lack of blood. It was as if a giant vacuum cleaner had gone through, in, and around the calf's carcass and sucked up every drop of its blood. We looked for even a speck of blood on the grass or on the animal's hide. Nothing. Not a drop. We looked at each other in stunned silence.

We ignored the temptation to scan the snow-covered perimeter of the pasture for something unworldly that still lurked out there and instead, with a tightness in our guts, went to work. We videotaped the crime scene and scanned the animal and the surrounding ground for magnetic and electric traces, for radio/microwave residue, and, for the hell of it, for nuclear radiation. No one knew who or what did this to the calf, so we thought we might as well check for everything.

I gently drew Tom aside. His normally ruddy face was the color of chalk. "Take me through it, step by step," I said. He tried cracking jokes about maybe a coyote with a scalpel had done it, but I could tell his heart was not into making jokes. He walked me to where he had tagged the calf just a few feet away, then walked me the couple of hundred yards to the west when about forty minutes later their snarling dog had given them the first hint that something was wrong.

"No noise," Tom said to himself, shaking his head in bewilderment. The blue heeler had still not shown up and it was now six hours later. "He just took off that way," Tom said, pointing west. "It's just not like him." He rubbed his eyes wearily. No more than forty minutes had passed in daylight, during which someone or something had entered the field, ripped the young calf away from its mother (she looked like she was still limping), drained it of blood, meat, and guts, and then carefully placed it on the grass. It just did not seem possible.

I looked at the calf's mother. She was standing a hundred feet away, her head down near the ground in a peculiar stance that exuded both aggression and fear. She was still breathing in a labored way. She never came closer as the veterinarian quickly stripped her calf's hide. An abrupt muttered exclamation caught my attention. "Look at the sharp cut," the vet exclaimed. "The ear was cut off with a knife or a scalpel." He was right.

Tom nodded and said, "That ear had a big yellow plastic tag." Looking closely, I saw the perfect incision where the cartilage and skin had been sliced. The whole ear had been cleanly removed to the skull. It was a beautiful, almost artistic, job. Even without lab pathology results, we all knew there was no way a predator or scavenger could have pulled this off. The cut was perfect and straight and had sliced cleanly through the tissue. One can easily spot the difference between the characteristic torn or ripped cattle hide of a predator or scavenger attack and the sharp cut of a scalpel or a knife when you look under a microscope. And here, even to the naked eye, this was obviously the work of a sharp instrument. The vet took samples just to make sure.

We watched silently as the vet finished the necropsy. He put the last bits of hide into labeled Ziploc bags for lab analysis. We carefully placed the torn femur into a plastic bag for forensic analysis. I looked around at the beautiful scenery and marveled that such an act of extreme violence could have taken place a few short hours ago in these pristine surroundings. Visibility through the trees and bushes on the perimeter of the large field was good, as the spring growth had not yet begun. Tom had told me this was only the second day of the thaw after winter. I scanned the entire perimeter over a hundred yards away. Nothing moved. To the north lay a ridge of reddish clay, probably weathered sandstone. The western perimeter of the pasture of the field was maybe three hundred yards away.

Before the light dimmed, we organized into groups and quartered the grass, looking for any footprints or vehicle tracks. The hard ground crunched underneath as we walked. The herd of all-black cattle kept their distance. They, too, were obviously disturbed. We agreed that the hard-packed grass could maybe have hidden some footprints but definitely not vehicle tracks. Tom watched us in bemused silence. He seemed relieved to have some company. Occasionally, we glanced nervously toward the perimeter to see if there was anything that shouldn't be there. We stopped our search as darkness fell.

As we returned to the science observation trailer less than a hundred yards from where the animal had been torn apart, we passed the three dog kennels. I could see all three animals huddled inside. One whimpered softly. "They haven't come out all day, even for food or water," Tom muttered. This was unusual behavior for three aggressive dogs that were used to roaming the land and fending off the coyotes and wild dogs that routinely passed through. And his other blue heeler still had not returned, now more than twelve hours later. In fact, Gorman would never see the animal again.

We spent the following day tracking the entire pasture, looking for any evidence, especially the big bright yellow tag from the calf's ear. We had summoned a professional tracker, who was due to arrive from Montana in a couple of days.

On the evening of March 12 we gathered in the observation trailer again to discuss the incident. The dogs had finally come out of their kennels but they were keeping close to us and the trailer. Even to my untrained eye, they had a hunted look. They were plainly still traumatized by whatever had passed through.

"Cattle are even more jumpy than yesterday," Tom observed. He was worried. The behavior of the dogs and the cattle suggested that something was lurking nearby. Tom talked repeatedly about the possibility of a predator on the loose. He simply couldn't fathom how somebody with equipment, including sharp instruments, could have pulled off this killing, with both him and his wife only a few hundred yards away. All of us knew from experience that Tom had uncannily sharp eyes and a keen sense of hearing.

Then, shortly after 11 P.M., all dogs suddenly started howling and barking hysterically. Tom, the physicist, and I leaped to our feet and ran out the door and into the cold clear night. We headed for Tom's old beat-up truck. On the driver's side, Tom had affixed a powerful spotlight that was easily maneuverable. He normally drove through his cattle at night during calving season, lighting up the animals as he drove. The powerful spot was excellent for seeing from a distance whether a cow or a calf was in trouble.

The truck engine roared as we bounced over the ruts. There were cattle here and there, some were milling nervously in the distance, some appeared oblivious. The animals seemed to be bunched toward the northern part of the giant pasture. On one of the turns, the headlights swept over a large black shadow standing in the southeastern tree line of the field. It was only a fleeting glimpse of what looked like a large cow, standing in the shadow beneath a large tree at the edge of the field, but far away from the rest of the herd.

"Must be in trouble," Tom grunted, as he swung the truck away from the main herd and headed south. The perimeter was about eighty yards away when the bouncing headlights suddenly picked up two large orbs of yellow light staring fixedly from the same large tree. "Did you see that?" two of us said in unison. Tom gunned the engine and, as the headlights danced crazily in front of us, we could now clearly see that the reflected yellow light came from the eyes of a huge animal probably twenty feet off the ground, perched in the tree.

"I'm not goin' to let him get another calf," Tom snarled as he suddenly ground the truck to a halt and reached behind for his rifle. We were no more than fifty yards from the large creature that lay motionless, almost casually, in the tree. The only indication of the beast's presence was the penetrating yellow light of the unblinking eyes as they stared fixedly back into the light.

This is unusual behavior, I thought as Tom steadied the rifle on the open door of the truck. Shouldn't it be running away? The rifle's sharp report rang out, and instantly, like a light being snapped off, the eyes disappeared.

"Got him," yelled Tom triumphantly. "I saw him fall to the ground." We scrambled back into the truck and Tom stopped about forty feet from the tree. There was no sign of the large creature under or near the tree. We split up and searched for any sight or sound of the wounded or dead beast. Thirty yards to my left Gorman suddenly yelled, "I see him." The shout was quickly followed by two loud reports from his rifle. "Got him at point blank," Gorman yelled as we jumped excitedly over the fence into the thick undergrowth. The snow crunched loudly under-foot as we stumbled around looking for some sign of it. We were cautious because a large, wounded animal was especially danger-ous at night.

I was still carrying my video camera as we warily looked all around, half expecting something to spring at us from the dark-ness. But there was utter silence.

"He jumped back here when I hit him," Tom said as he scratched his head in puzzlement. "I was no more than forty feet away. Where'd he go? That sucker must have weighed four hun-dred pounds."

We began quartering the area, looking at the snow for tracks. Something that big should have left an obvious trail in the snow and there should have been blood. It was then that I saw it—a single, obvious oval track about six inches in diameter embed-ded deeply in the patch of snow. I yelled at the other two, who came running. I shone the flashlight, and there it was. It looked unusual: a single large print in the snow with two sharp claws protruding from the rear of the mark going a couple of inches deeper. It almost looked like a bird of prey, maybe a raptor print, but huge and, from the depth of the print, from a very heavy creature. I began videotaping, as the physicist unsteadily held the flashlight while looking over his shoulder, waiting for a large wounded animal to charge us.

Minutes ticked by as we searched in vain for a companion print. We found another one in an area of unmelted snow about twenty feet from the first, but nothing else. We listened care-fully for any movement in the undergrowth as we searched. There was an eerie silence in the area, broken only by the distant bellowing of the cows over on the other side of the pasture. They were still deeply disturbed by the commotion we had gen-erated and were not coming over to investigate.

Almost two hours later, we gave up, exhausted, cold, and still slightly jangled from the brief burst of adrenaline. Again there was an unspoken shudder at the idea of sleeping in the trailer a mere hundred yards from where the bizarre creature or creatures had apparently vanished into thin air, having been shot with a high-powered rifle. It was time to call an end to another busy day on the Skinwalker Ranch.

The next day we compared notes and debriefed Tom on what had happened. He swore that there had been two large animals, one in the tree and the other on the ground. He also swore that he had hit both animals, the first with a single bullet, the second with two bullets. Tom was an expert marksman, able to kill a coyote at five hundred yards, and he had shot both of these large animals at close range. He described the second animal that crouched in the undergrowth as huge, heavily muscled, and looking like a four-hundred-pound wild dog. The animal had been staring at him and had tensed just before springing when Tom had shot it twice from no more than forty feet. Yet no signs of blood and two confusing distant single tracks in the snow added to, rather than solved, the mystery.

The incidents of March 10–12, 1997, were very difficult to explain. They were certainly verification of unusual or anomalous activity, but did they fit the criteria that NIDS was looking for? Those criteria initially had been defined as the verification of events first described by Gorman, the witnessing of unidentified flying objects by the scientists, and their capture on camera or video images. The bizarre killing of the calf and the potentially deadly nighttime encounter with phantom creatures that were shot but left no blood and few tracks did not appear to fulfill the criteria of anything that could be reported at a scientific meeting. Nor could this kind of incident be written up for a peer-reviewed science journal. In fact, beyond the videotaped track in the snow, there was no physical evidence that this incident had ever happened. And at a subsequent Science Advisory Board meeting, the board made it abundantly clear that these events, in the absence of physical evidence, did not constitute verification of anything.

Little did we know that these unexpected events would become part of an increasingly frustrating pattern of transient, difficult-to-interpret, but frightening events and phenomena that would never again be repeated in our presence.

CHAPTER 16
The Bulls

After the events of March involving the uncanny mutilation of the calf followed by the equally bizarre appearance of the phantom "predators," things were relatively calm at the ranch. But Tom mentioned at the very beginning of April that something just did not feel right. He felt this way whenever the phenomenon was close by. It manifested itself as an unaccountable feeling of oppression, a feeling that someone or something was watching, waiting to act. In short, he told us, it was a very uncomfortable sensation, like something bad was about to happen. And it did.

On April 2, Tom called me to report that yet another calf had mysteriously disappeared from the well-fenced property. This was the fifth animal to "disappear" in 1997 without explanation. At the same time, Tom observed that the long-term ranch dog had disappeared and hadn't been seen for a couple of days. Again, he found it unbelievable that five animals could disappear without any signs or tracks. His frustration level mounted. Animals were disappearing under his watch without explanation and without any physical clue that might assist the NIDS investigation.

Tom's bizarre story led me to organize an expedition to the ranch the following day. The team, which included a very accomplished and seasoned investigator from Canada, arrived the following afternoon. It appeared we had just missed another incident by a couple of hours. As soon as I set foot on the property, I could feel that the atmosphere was laden with a thick creepy miasma. I noticed the dogs huddled in their kennels, looking very intimidated.

Tom and his wife were very shaken. This had been one of the very rare occasions when Ellen had ventured onto the ranch since leaving the previous August. We repaired to the central command post to hear his story. Two major episodes had occurred within less than twenty-four hours.

On the afternoon of April 2, Tom and Ellen had set off toward the west end of the ranch on a routine mission to spot and count the animals. As they passed the bull enclosure, both of them looked fondly and proudly at the four burly bulls in the corral. They truly were magnificent beasts, two each of pure black Simmental and Black Angus, each weighing more than two thousand pounds. With muscles rippling healthily beneath the shiny black coats that perfectly reflected the setting afternoon sun, the animals made the Gormans proud. Ellen said wistfully, "I would go out of my mind if I lost any of those animals." Tom nodded in agreement as they drove west on the narrow dirt track past the corral.

Forty-five minutes later they drove back. All the animals seemed to be accounted for, yet they could not shake that nagging feeling of unease. An unnatural calm hung over the property, broken only by the sound of the truck engine. Abruptly Ellen screamed and pointed out the windshield. Tom hit the brakes, fearing he was about to run over something. He followed her finger and gasped. The corral was empty. Tom's stomach knotted. Each of those four registered bulls was worth thousands of dollars. They were irreplaceable. Tom looked into Ellen's tear-stained face. They stopped the truck by the empty corral, and he got out to search for some evidence that the four magnificent animals could have left behind. Tom's knees felt weak. There was no sound as he walked around the corral.

Tom walked around looking at the footprints in the corral. The animals had been there only forty-five minutes ago. Ellen was sobbing in the truck. His search meandered over to an old small white trailer located at the west end of the corral. There was no entrance to the trailer from the corral except a door that was tightly locked and hadn't been opened in years. As he passed the trailer he glanced in. Tom froze. All four animals were standing silently, crammed into the tiny space. They seemed frozen hypnotically and appeared to be barely conscious. Tom, with relief flooding through his veins, yelled loudly for Ellen. At the same time he banged forcefully on the side of the metal trailer. The noise seemed to break the silent spell. Instantly, all four animals appeared to wake up. They began kicking and bellowing to get out of the narrow, confined space. Within seconds the four huge animals went berserk and devastated the interior of the trailer. Finally, a metal door was kicked out and instantly all four animals tumbled blindly out the broken door and began stampeding in a panic.

It took Tom and Ellen several hours of cajoling and skillful cowboy tactics to get the four panicked animals safely back into their corral. Night had fallen by the time they were finished. Silently they drove the twenty-five miles to their home. Both were physically and emotionally drained.

As soon as Tom had finished his story, we tumbled out of the command and control headquarters and, with cameras and instruments in hand, ran the twenty yards to the corral where the four huge bulls stood. The animals seemed wary, almost jumpy. I walked over to the tiny trailer and noticed the locked door, bolted with a piece of barbed wire through the bolt. Inside the devastated trailer I carefully examined the inside of the door. This door was the only possible entrance point from the corral to the cramped interior of the trailer. I noticed the extensive cobwebs on the inside of the door. There was no evidence that the door could have been opened without disturbing the layers of cobwebs. As Tom had said, the inside of the trailer was thoroughly trashed. Bull dung was everywhere. The stench was strong. I emerged quickly from the foul-smelling, cramped space and marveled that such huge animals could even fit inside such a small area.

We then began a systematic sweep of the corral and surrounding area using the small portable instruments that were part of our tools of the trade. Right away we realized that the metal bars of the enclosure, especially in the region near the trailer, were highly magnetized. The needle of the magnetic field detector went off the scale when we applied the instruments to the bars of the corral nearest the trailer, but the magnetic field was absent on the other side of the enclosure. Something had apparently magnetized the bars of the corral nearest to where the four animals had been found. We repeated the measurements every few minutes.

Our team also began taking multiple photographs and we videotaped the entire area including the wary bulls. They gazed suspiciously at us as we roamed the area, systematically checking for any additional unusual radiation. There was nothing apart from the strongly magnetized metal bars of the corral, which diminished as time went on. Forty-eight hours later, the magnetic field was barely detectable. We stopped our investigation as the sun was setting. On the way back to the command and control center, we noticed that the dogs were still hiding in their kennels. One blue heeler was whimpering softly.

As we unpacked the instruments we had brought with us,

Tom began relating the second incident that had happened only a few hours earlier. Tom and Ellen had gone to the west end of the property to check on the herd that was grazing. After the inexplicable incident the previous day in which their bulls had briefly disappeared, they were on the lookout for anything unusual. All seemed quiet as they parked their truck in the pasture. They watched dozens of well cared for animals grazing nearby.

Fifty yards away, one of the cows wandered casually toward a bright blue salt lick that lay between two straight lines of poplar trees that jutted a hundred feet into the air. Tom watched as the animal suddenly stopped dead in its tracks and, head lowered, began to back away from the salt. Tom gently nudged Ellen and pointed. Ellen jumped nervously and turned her head toward the animal. She, too, watched as the cow, transfixed with fear, was literally backing away from something invisible as fast as its legs could move. Tom reached through the window of his truck and grabbed his compass. He had learned a long time ago that strange events caused the compasses to act strangely. The frightened cow had now backed thirty yards away and suddenly turned tail and stampeded toward the eastern end of the pasture.

Tom then watched as the herd of cows suddenly split in two in Red Sea fashion, as if something invisible was now in the middle of the herd of cattle. Dozens of animals ran west, while the remainder stampeded east. Tom glanced briefly at his compass and saw that the needle was pointed to the invisible something in their midst. Whatever was dispersing the cows was invisible even in the noon sun, yet it was exuding a powerful magnetic field that was detectable on Tom's compass. Tom and Ellen could follow the mysterious, invisible object as it made its way through the herd of animals, as the panicky animals were constantly stampeding away from it. Whatever it was, it was heading south, as did the needle of Tom's compass.

Sitting in his seat in the command and control center, Tom told us, "The needle of the compass stayed locked onto that thing for exactly eleven minutes." So there had been two mag-netic field effects at different ends of the ranch in less than twenty-four hours. "That is the first vague pattern that I have seen so far," I commented.

The next two days were uneventful as the team divided up and began systematically mapping the ranch for the magnetic and electric fields that appeared to be diagnostic of the presence of the phenomenon. But apart from the magnetic field that radi-ated from the bars of the western part of the corral near the small white trailer for a couple of days, we saw nothing.

Mapping the south side of the ranch, with its acres of dense Russian olives and thick brush, was slow and tough going. Tom took us to a region in the brush where he had come across un-usual tracks a few days previously. The tracks were perfectly round and appeared to be from some mechanical object rather than from a known animal. The team took plaster casts of the mysterious tracks.

Then as we were quartering the ground about twenty yards away, an overpowering stench of musk assailed my nostrils. I had spent enough time on the ranch to know the difference among fox, skunk, and this particular odor. It was very strong and ap-peared localized. I carefully looked around. I could see another team member, the experienced Canadian investigator, about fifty yards away making noise as he searched the dense brush for tracks.

Suddenly, the hairs on the back of my neck stood up. A chill ran down my spine. Something was close by and watching me. The locus of the strong musk odor was coming from my right, and I slowly turned my head in that direction. But as my eyes quartered the area, I could see nothing. Since it was broad day-light, I could only conclude that I was either imagining things, or else there was something well camouflaged nearby. This was one of the very few times, in the hundreds of days that I spent on the property, that I ever felt physically threatened. I gritted my teeth and continued the mapping. Gradually, the intensity of the musklike stench began to recede, and at the same time I began to sense that whatever was close by had left.

Later the Canadian investigator told me that he too had suddenly become aware of being watched. He could not explain it. He had also smelled the strong musk odor. Tom, who was a lifelong hunter and was familiar with the scent of all wildlife in the area, swore the smell was associated with the "phenomenon," whatever it was. That smell came back to haunt us repeatedly throughout our research program on that ranch, appearing suddenly without warning, sometimes in a highly localized way, sometimes right inside our command and control center. Whatever it was, its invisibility sure gave it an advantage over us.

At about this time we began to move around the ranch accompanied by dogs that served as biosensors. The dogs were much more adept than we were at picking up traces of the phenomenon. Throughout that period in early April, the dogs continued to act as if something was wrong. At the same time I noticed that eerie sensation of an unusual presence in the environment. It was nothing I could put my finger on, a sustained uneasiness and an inner expectation that all hell could break loose at any time. Was I imagining things?

We were in the command and control center planning our midnight watch on April 7 when the phone rang. It was Julia Sanchez, our neighbor who lived a couple of hundred yards east of us in a trailer. Her excited voice told us to get outside *now*. Three UFOs were heading our way. Three of us stumbled quickly out of our trailer and searched frantically around the sky. An eerie silence hung over the property, but there was no sign of low-flying UFOs.

Sanchez later told us that the family had seen UFOs on several occasions in the past few days and this was only the latest incident. We searched the night sky for half an hour with night-vision binoculars but saw nothing unusual. We spent hours out there looking at the occasional meteor. The dogs were beginning to act normally again. I began to wonder if the phenomenon, whatever it was, was moving on. The air itself seemed less threatening.

A day and half later, the rain began. The downpour began on the afternoon of April 9 and seemed to intensify as the sun set. In northeastern Utah, a rain downpour can be pretty severe. Visibility drops significantly and the ground itself becomes an oily, dangerous quicksand where a car can slide quickly into a ditch and get stuck. At midnight, as we tried to prevent our fishtailing vehicle from sliding into the four-foot-deep canal, I spotted a dim light at about two hundred yards to the south. The team had been scanning the area for hours. Now, suddenly, this light had appeared out of nowhere. We knew that area comprised very thick brush, and as the rain hammered down on the vehicle roof, I decided to reconnoiter up to higher ground to see if we could get a fix on the object.

We maneuvered the sliding vehicle away from the canal and drove a couple of miles off the property to a ridge where we could see the entire ranch from above. The rain appeared to be getting even heavier as we climbed out of the vehicle. Within minutes we were soaked as we scanned the terrain below for any sign of the mysterious light. I trained the generation-three night-vision binoculars, and instantly a large bright light on the ground became apparent. I moved the binoculars away from my eyes and I could see the dimly lit object only as a barely discernible faint dirty yellow light. Through the infrared night-vision binoculars, the object looked like a very bright, large white beacon resting on the ground about a mile below us. This curious discrepancy would become more meaningful later on in the summer.

I knew we were insufficiently equipped to negotiate the dangerous terrain below in what was perhaps the heaviest rainstorm I had ever experienced, so we contented ourselves by mapping the location of the object so we could return there when the rain diminished slightly. At 3 A.M. we called it a day and returned to the command and control center. The next day, when the rain had subsided, we spent eight hours searching in the deep brush for evidence of the object but found nothing.

The following morning we were awakened when Tom and Ellen arrived to take a look at the cattle. It had been a couple of days since they had visited. They needed a break after those disturbing early April experiences. As the Gormans took seats in the command and control center, one of the team members casually switched on the portable magnetic field detector to check for battery power. Instantly, the needle jumped off the screen. The disturbance centered on Ellen Gorman. We had previously carefully calibrated the inside of the command center to make sure we knew which appliances, instruments, and areas produced magnetic fields. We had a recognizable and consistent baseline magnetic field "footprint" for the inside of the command center. This dramatic surge in magnetic field intensity was inconsistent with anything we had previously seen. It was strongest about two feet from Ellen.

When she left the trailer, the field had disappeared, and a few minutes later, when she re-entered, there was no field present. Although we repeatedly checked the same area of the command center for magnetic fields, no recurrence of the intense magnetic field ever happened. We checked the instrument itself and compared with other instruments. It was working normally. We ruled out instrument malfunction.

Again, a mysterious magnetic anomaly had occurred that was strong but frustratingly brief. But we had still little or no data that was reproducible or robust, certainly nothing that would constitute strong evidence of anomalies or that could be reported to the Science Advisory Board.

Following the magnetic surge, events appeared to return to normal on the ranch. The dogs began running and chasing around the property, as they had not done for ten days. The cattle settled down and were less likely to stampede in response to unseen stimuli. The previous ten days had seen multiple fleeting events that seemed to indicate something in the environment but nothing that we could measure or even see. By the end of April, it was becoming frustratingly clear that the well-laid NIDS phase-one plans to initially obtain robust personal sightings by the scientific team were not coming to fruition. Instead, the team was encountering tantalizingly brief anomalies, few of which were repeated and even fewer were measurable. It was more than six months since the NIDS team had initially deployed with the expectation that phase one would be complete within a few weeks.

In May 1997, the NIDS team had a series of enclosures built on the ranch to house the dogs. The dogs had shown consistent patterns of being alert to whatever was in the environment and had always been able to warn us by their barking. We had learned to watch the dogs closely. In most cases, aberrations in their behavior signaled the beginning of some kind of unusual activity.

The large wire enclosures measured about eighty feet by twenty feet by fifteen feet high, and at each end, a wooden viewing platform was constructed so that team members could see the area from a vantage point twenty to twenty-five feet away. They looked like large deer-hunting stands, only much more robust. Three structures were positioned at the east end, middle, and west end of the property.

The enclosures were struck by either some very skilled intruders or by what seemed to be a torrent of "poltergeistlike" activity. Persons unknown would routinely open the doors, padlocks would vanish, and the dogs would escape by unknown means. Even the strong wiring that secured the inner doors would vanish mysteriously. It would take considerable effort for someone to mount this kind of consistent interference and damage without leaving tracks or any signs of intrusion.

Throughout this period we deployed teams on night watches to try to catch the perpetrators, but without success. The mysterious interference continued at a frenzied pace throughout the month, but nobody was ever caught, in spite of multiple stakeouts by trained investigators. We never found any direct evidence of intruders in the nine dozen instances of interference with the three enclosures. And throughout this period, the dogs were very jumpy. By the end of May, when the interference with the enclosures had finally died down and the dogs' behavior began to return to normal, the activity on the ranch took a dramatic shift.

CHAPTER 17

Encounters

It was darker than I can ever remember. The only light came from the stars and some very distant yard lights that you could see if you moved your head back and forth so the trees no longer blocked them. It was close to midnight and still warm. We were thankful there were no mosquitoes that night; sometimes in early June, they can start swarming. I remember watching a colleague get out of his vehicle in Utah when the mosquitoes were bad. Within minutes his blue jeans had turned a seething mass of gray.

The two dogs we had brought along as biosensors were running around sampling the many smells of wildlife. Raccoons sauntered through frequently and coyotes could be heard regularly, though not on this night.

We had just arrived in Utah from Las Vegas, and this was the first night of watching. "Position yourselves exactly in front of the left-hand window in the old homestead," the Canadian investigator had advised us. The week before he had shot some beautiful infrared film of a very eerie bright light that had appeared from that vantage point. Though he never actually saw the light, it had shown up on infrared film. But only one Kodak Ektachrome frame showed the intense light in a sequence of four. So we knew if we were going to see anything in this location, it probably was going to be transient.

The NIDS physicist and I gazed out on the field. Nothing moved. Even the dogs were silent. My colleague stood maybe ten feet to my right. He had night-vision binoculars. Down on the other end of the property stood two more intrepid investigators, also with night-vision binoculars. When you clicked them on, everything became very clear, though in a bizarre shade of green.

Then it appeared without warning, no more than seventy-five yards to our left. A silent, brightly lit sphere of bluish-white light about the size of a basketball hovered, moving slightly as if swaying gently. The dogs, which were behind us, seemed to notice its abrupt appearance. The object was not more than fifteen feet off the ground. It appeared to be bobbing slightly and was bright enough that I could see the grass lit up below it. There were no obstructions in our line of sight. The thing was definitely within our same small pasture. I could not hear any wildlife sounds; it was as if a blanket of silence had descended on the area. We stared and, just as abruptly, it was gone—just as I had begun to train my camera on the object. It was as if somebody had thrown a switch. The dogs did not move.

Immediately, we clicked on the powerful Maxa-Beam. According to the literature, you can read a newspaper at night from a mile away with the light of the Maxa-Beam. Shining it in somebody's eyes would bleach their retinas, so we were always careful. The powerful instrument is a favorite of military and law enforcement personnel.

Nothing moved as the entire pasture was lit up in bright white light. Slowly, we scanned the whole pasture. Nothing appeared out of the ordinary. We walked over to the spot where the object had appeared and, just as suddenly, had disappeared. The dogs stuck close by our sides. They were not in the mood for playing. We searched the area quickly for a few minutes but could find nothing. Slowly and warily, we walked back to our original position.

My colleague was scanning the perimeters of the lush, tree-lined pasture with a pair of generation-three ITT night-vision binoculars. These binoculars amplify ambient light both in the visible and, to a large extent, in the infrared. I was readying the manual camera with black-and-white infrared film when he exclaimed, "Jesus!" He was looking through the night-vision binoculars, directly at the tree line no more than two hundred feet in front of us. All of a sudden he said, "There's a huge black thing in the trees just in front of us and it is moving north."

That certainly got my attention—and the dogs' as well. Both animals had taken up positions directly behind us, jammed into the backs of our legs and gazing fixedly ahead where my colleague was looking. I pointed my manual camera in the direction he was looking and began a series of long-exposure shots.

"It is big, and I'm not sure if it is in the trees or behind the trees," he said. "It is blocking out the stars." I kept the shutter open and began counting out about twenty seconds between opening and closing the shutter. The camera was mine, probably forty years old, but capable of taking excellent shots. Experience told us to avoid the high-tech "idiot-proof" cameras whose electronics had too often failed at the crucial moment.

Every time I looked up to see what my colleague was reporting, I could see nothing except the dark shadows of the tree line directly in front of me. Without the advantage of the amplification of the low-level ambient light afforded by the night-vision technology, I was looking for a black something against a black background. I decided to focus only on my camera work.

"It's still moving," he was muttering. Then, all of a sudden,

"It's got me," he yelled. "It's saying, 'We are watching you.'" Then there was silence. I kept taking increasingly longer exposures to try to catch whatever he was talking about. I could not see what was causing him such intense anxiety. My colleague's frantic actions and tone of voice increased my adrenaline. I knew that if something out there in the dead of night wanted to harm us, we were sitting ducks. Then he said, "It's getting smaller." Then, "It's gone." Over and over, he kept muttering, "Jesus Christ. Jesus Christ."

I asked him what had happened. He was still shaken. "Something big was in the trees just in front of us, it blotted out all the stars through the binoculars," he declared. "It took control of my mind. It told me it was watching us." He sounded very confused and bewildered. I had worked closely with him on numerous occasions and knew he was not prone to sudden flights of imagination. His distinguished academic career had not prepared him for anything like this.

We stayed in the area about another forty-five minutes. At one point he went inside the old homestead and almost instantly started hollering. He had disturbed a sleeping bird that had suddenly flapped its wings in his face. I realized he was probably more upset than he was admitting. So we began to pack up the equipment. It was time to call it a night. We were silent as we trudged back the mile to the sleeping quarters. I kept thinking about the weird experience. The dogs. The invisible thing. The infrared. I went to bed that night thoroughly puzzled.

Throughout that summer, the NIDS team spent many long hours in the same area where the interaction had taken place with a telepathically hostile voice. We played cat-and-mouse games through the months of July and August with multiple different lights that appeared for a second or two, then vanished, then reappeared several hundred yards away. It was if something intelligent was leading us on a dance as we rushed silently from spot to spot, always just slightly behind the fleeting orbs of light.

Sometimes, they appeared a few yards away in the middle of trees, or else they were a long way away. But we were never able to capture these nighttime game players on film. It was an exhilarating but ultimately frustrating summer.

Then at the end of August 1997, on one of the few occasions when the NIDS team was not on the ranch, Tom encountered something truly bizarre. The cattle had been skittish and nervous for days. At about four o'clock in the afternoon, the cows were startled by something and panicked. They stampeded south, breaking through the fence line that bordered the property. Over the summer, Tom had become weary from the cattle destroying the barbed-wire fences. Of course an animal has to be quite panicked to break into barbed-wire fencing. The act usually draws a lot of blood and leaves lacerations, some of which may later become infected.

Tom saddled up his horse and headed off after the cattle. They had broken into a neighbor's alfalfa field and there was a danger that some of the animals would die from bloat. Eating growing alfalfa at this time of the year was very dangerous for cattle. The plants balloon in the rumen and cause such bloating that it sometimes kills the cow. Occasionally, a veterinarian may be called on to puncture the rumen with a sharp instrument so that the built-up gas can escape. This drastic action usually saves the cow's life.

Tom hurried back to the property and called his teenage son for help. There was no way he could herd the three dozen animals out of the juicy alfalfa without some help. Over the years, Tad had become familiar with the ritual of herding panicked cattle back through the broken fence lines. The setting sun cast a beautiful golden glow over the restless animals as they twisted and turned to evade Tom and Tad on horseback. Tom noticed that the animals seemed fearful about going back to the ranch. For two hours, he and his son gradually moved the animals a few hundred yards to the north. Then something would happen, a ripple of panic would run through the herd, and they would charge blindly back south for several hundred yards and stop, breathing heavily while looking fearfully back the way they had come.

By midnight, Tom was becoming frustrated. In an act of desperation he tried a different tack. Instead of heading straight north, he and Tad drove the animals east first and then tried to take them north at a different point. He hoped to avoid the area that seemed to be driving the animals crazy with fear. The new path took them within fifty yards of a creek that bubbled noisily below a drop of about fifteen feet.

As he rode in the direction of the creek, a golf-ball-sized blood-red ball abruptly flew into view and came directly at him from the creek. The object just missed the horse's head as it flew past, and Tom felt a stab of fear. The horse reared up and began to run. Tom had to fight to calm her down. Off in the distance, where Tad rode among the milling cattle, several were bellowing loudly in a crazed way. Tom looked to his left and saw one of his prized bulls chasing its tail and bellowing. The animal was plainly mad with fear. Then Tom saw the small red ball darting around the bull's head. Tom called Tad over and quietly asked him if he had seen anything. The boy responded by nodding dumbly. Tom could see the fear in his son's eyes.

"Let's give it another try," Tom muttered. Suddenly, out of nowhere, another blood-red golf ball came straight at his horse. Tom lost control and the horse took off in full flight, stampeding crazily toward a canyon. Tom knew the animal was not going to stop. Ten yards before the horse plunged over the edge of the canyon, Tom threw himself off the animal and landed heavily on the ground, only a few feet from the edge. Miraculously, the horse landed twenty feet below without obvious injury and began making the arduous climb back to the canyon lip. Tom, badly shaken, spent ten minutes trying to coax the sweating horse into allowing him to remount.

Out of the corner of his eye, Tom noticed two of the small red objects moving among the stampeding cattle. They seemed to be herding the animals in the direction of the creek. Tom and Tad saw what was happening, but it was too late. They tried to intervene as the cows and their calves stampeded over the fifteen-foot drop into the creek below. When the melee had settled, Tom could see several cows and calves lying prone. They had been trampled, and a couple had been injured in the fall into the creek. There was no sign of the red balls.

The other animals had clambered up the other side of the creek and were making their way, still bellowing but slightly less panicked, back toward the ranch. Tom called out to Tad and together they began to herd the now pliant animals toward safety. After they were safely back on the property, Tom returned to the creek to survey the devastation.

One cow was obviously in deep distress, a couple of calves had broken legs, and a fourth cow lay on her side moaning and shivering. The loss was less than he had anticipated. In the distance, he saw the headlights of Ellen's truck as she pulled onto the property. Wearily, Tom turned his horse back. He saw it was past 2 A.M. The high-intensity struggle with the blood-red objects had lasted about seven hours. Ellen parked the vehicle under one of the surveillance cameras and Tom wearily climbed in. He felt exhausted. The moon had climbed over the horizon and was bathing the pasture in a dim light. In the distance, Tom saw Tad climbing over the fence line from the adjacent pasture, having tried to help the stricken animals in the creek.

Suddenly, Ellen nudged him and pointed. Tom rubbed his eyes and followed his wife's outstretched arm. Hovering silently about ten feet over his son's head was a blood-red ball. She screamed at Tad, but he was unaware of what was above him. Suddenly, the small red object took off toward them, flying low to the ground and directly into the glare of the truck's headlamps. Passing only a few feet from the truck and accelerating as it gained height, the object quickly vanished into the westerly sky. Tad approached the truck, looking puzzled. He admitted that he had an eerie feeling that he was "being watched" just before his mother screamed at him. Otherwise, there were no ill effects.

All three sat silently in the truck as the clock ticked past 2:30. Tom then climbed out to check some irrigation gates and said he would meet Ellen at the exit to the property. She nodded and gunned the engine. It took Tom about five minutes to change the irrigation ditch water flow, and he took the shortcut near the old ranch house. Just beyond the homestead, Ellen was waiting for him in the truck.

As he climbed in, Ellen silently nudged him. Again he followed her trembling finger. Just adjacent to the old homestead, a bright blue ball hovered maybe a dozen feet above the ground. It was a familiar sight, but they had not seen one of these objects in several months. Larger than a baseball but smaller than a basketball, it emitted an intense electric blue light onto the nearby building as it hovered silently. It seemed to be watching them. Tom and Ellen watched back silently, wondering what other surprises were in store for them. Suddenly, the blue object darted behind the building, out of their sight. No one said anything during the twenty-five-mile journey back to their home.

It had been another busy day at the ranch. One cow aborted, another calf died, and the others recovered slowly from the injuries inflicted that night. All the animals had been traumatized by the mysterious golf-ball-sized objects. Tom had never seen these objects before and hasn't seen them since.

That morning Tom called the NIDS team back to the ranch.

CHAPTER 18

Mystery

The late physicist Richard Feynman, one of the towering figures of twentieth-century science, was publicly skeptical about unidentified flying objects. During a lecture at the University of Washington in 1963, Feynman summed up his feelings in a few biting paragraphs about what he called "This Unscientific Age."

"If we come to the problem of flying saucers . . . we have the difficulty that almost everybody who observes flying saucers sees something different," Feynman remarked. "Orange balls of light, blue spheres which bounce on the floor, gray fogs which disappear, gossamer-like streams which evaporate into the air . . . round flat things out of which objects come with funny shapes that are something like a human being. Just think a few minutes about the variety of life that there is. And then you see that the thing that comes out of the flying saucer isn't going to be anything like what anybody describes."

Feynman didn't know it at the time, but his brief description of assorted UFOs that had been reported by 1963 could almost qualify as a laundry list of the varied and mysterious objects seen three decades later on or near the Gorman ranch. Orange balls, blue spheres, disappearing fogs, round flat things from which humanlike forms emerge, and many other unusual aerial phenomena were witnessed by the Gormans, by their immediate neighbors, by much of the population of the Uinta Basin and, eventually, by the NIDS scientists.

The most common UFOs seen on or near the Utah ranch were balls of light, of various sizes, colors, and intensity. Some were described as resembling yellow headlamps, others as glowing orbs that flitted across the pastures, through the trees, and over the ridge, operating under what seemed like intelligent control. A few of these incidents at the ranch were captured on videotape. Amorphous balls of light are probably the most commonly reported UFOs worldwide. Without question, such sightings are often the result of the misidentification of planets, meteors, and other natural phenomena. But as the Gormans learned all too well, natural phenomena cannot account for a lot of what they saw.

Odd balls of light, similar to what the Gormans reported, were the first unidentified objects to be seen during the modern UFO era. In the latter years of World War II, Allied pilots frequently spotted and even photographed balls of light that trailed their planes. The Allies suspected that these objects, which came to be known as "foo fighters," might be German secret weapons. But the Germans also saw the foo fighters and thought they were ours.

In the decades that followed, secret military studies were conducted concerning the UFO mystery. These once-classified military files are filled with cases that sound nearly identical to the objects seen at the ranch. One witness was a military pilot named G. F. Gorman. While flying his F-51 near Fargo, North Dakota, in 1948, Lieutenant Gorman encountered an intense white light about a foot in diameter that made dramatic turns and maneuvers during a thirty-minute aerial dogfight of sorts before it finally disappeared with a burst of "stupendous speed." Several witnesses on the ground watched the encounter. The balls of light the Gormans witnessed seemed capable of similar maneuvers.

Balls of light figure prominently in some of the most famous and best-documented UFO cases. In 1951, thousands of witnesses in Lubbock, Texas, saw a formation of blue lights as they traversed the sky on several nights over a two-week period. Military radar tracked the lights and determined they were traveling nine hundred miles an hour at an elevation of thirteen thousand feet. Several photographs were taken of the lights. The most prominent UFO debunker of the day, astronomer Donald Menzel of Harvard, alternately explained the lights as being the reflection of streetlights, headlights, or house lights against an unseen "rippling layer of fine haze," although Menzel was never able to explain how reflected headlights could be picked up on a military radar. Other theories suggested that the lights were caused by reflections of light from birds or moths.

UFO skeptics have long relied on obscure atmospheric or weather phenomena to explain away otherwise inexplicable aerial mysteries, even when those explanations simply don't fit. For example, in the summer of 1952, unidentified balls of light were seen flying over Washington, DC, on consecutive weekends. The lights were seen from the ground and the air, were detected on military and civilian radars, were photographed, and were repeatedly pursued by military jets. The explanation offered by the Air Force was that the false radar sightings had been prompted by a temperature inversion in the area, a theory that simply did not fit the facts surrounding the sightings and that was publicly disputed by the military's own radar operators and pilots.

Another oft-cited explanation for unidentified lights is that of plasmas or ball lightning, rare and little-understood phenomena that have been known to generate small clouds of electrified air that could conceivably be mistaken for UFOs. Prominent UFO critic Phillip Klass, a well-known aviation writer, championed the plasma theory. Although the theory could theoretically account for some sightings of balls of light, it failed to gain much acceptance, even among plasma physicists who were generally hostile to the UFO topic. A pro-UFO atmospheric physicist named James McDonald systematically demolished the ball lightning explanation for the simple reason that plasmas can't do the things that UFOs do. Ball lightning generally lasts only seconds or fractions of seconds, does not travel at high altitudes over long distances, and cannot possibly account for those balls of light that are seen in cloudless, stormfree skies.

Yet another suggested explanation for unidentified balls of light is earthlights, which are electrical discharges in the atmosphere generated by geological forces in the earth. As with ball lightning, the earthlight phenomenon is real, but it cannot account for sightings of longer duration in which UFOs perform controlled maneuvers or interact with the observers. At the Utah ranch, the NIDS scientists persistently searched for natural explanations for the sightings. Their research included a detailed analysis of geological data and weather patterns for the area. But we found no natural explanation for the things seen by the many witnesses.

The most intriguing incidents at the Gorman ranch most certainly could not be explained as the misidentification of amorphous, random bursts of natural light sources. The facts simply do not fit. By any stretch of the imagination, ball lightning and earthlights could not reasonably account for those UFOs that were described by eyewitnesses as structured, metallic craft, including saucers and discs, dozens of which have been seen in the vicinity of the ranch over the past several decades.

Although the Gorman family saw many unidentified objects flying above their ranch, few would qualify as classic "flying saucers." However, saucers and discs have been seen on the property and on neighboring ranches as well. Mr. Gonsalez, a neighbor, told us of seeing a UFO shaped "like a Mexican hat" that flew over his home. (Saucers and discs have often been compared to sombreros because of domes that are spotted on the tops of some discs.) Gonsalez and his family also saw a silvery disc that flew directly into the rocky crags of Skinwalker Ridge and seemingly was absorbed by the ridge itself. The craft merged with the earth with no noticeable disturbance or impact.

Another sighting occurred fourteen years before the Gormans arrived in the area. Reporter Zack Van Eyck of the *Deseret News* interviewed Salt Lake City resident Dean Derhak, whose uncle owned one of the properties that border the ranch that would later be purchased by the Gormans. While riding a horse on his uncle's land in 1980, Derhak says he saw a "silver sphere" sitting on the ground of the neighboring parcel.

"It was fairly big, about 30 to 40 feet wide. It looked like a bowl upside down," Derhak told Van Eyck. "It scared me and I took off."

These sightings by neighbors are consistent with the widespread belief among area residents that the family that owned the property prior to the Gormans was also very familiar with assorted unexplained phenomena. Some residents expressed the opinion that the prior owners even developed some sort of live-and-let-live relationship with whatever inhabits the property.

Other objects seen at the ranch resemble some well-documented sightings of objects in other parts of the world. Tom Gorman says one of the objects he saw most frequently was something he described as round, orange, and elongated, like a setting sun, but it moved like no sunset he had ever seen before. Something large, round, and orange was seen in 1975 hovering above the Minuteman missile silos at Malmstrom Air Force Base in Montana. Government documents obtained through the Freedom of Information Act reveal that when a security violation alert was issued, an elite Sabotage Alert Team (SAT) was dispatched to the scene. SAT members witnessed a large, glowing, orange disc floating over the top secret facility, so bright that it illuminated the missile silos below. Fighter jets were scrambled in response to the mysterious intrusion, but the orange ball flew straight up until it disappeared from radar at two hundred thousand feet. The fighters were understandably unable to follow it. A subsequent investigation revealed a more shocking development—the launch codes of the missiles had been inexplicably changed during this disturbing overflight. Equally troubling are the Pentagon documents that acknowledge that four other nuclear missile bases along the U.S.-Canadian border had similar intrusions by mystery lights within a few weeks of the appearance of an orange "sun" at Malmstrom.

Tom and Ellen Gorman also reported separate sightings of a silent, hovering, triangular craft that projected multicolored lights from its black frame. They could discern no known propulsion system and wondered if the triangles might be something akin to a Stealth fighter or B-2, both of which are black and triangular. Tom said the triangle crept silently across the pasture as if it was searching for something. Ellen saw a black triangle that kept pace twenty feet above her car. Without overstating the obvious, it should be pointed out that both the F-117 and the B-2 produce a considerable amount of noise during flight, although rumors of classified silent versions of both aircraft abound. No known public versions of either aircraft are capable of simply floating in silence over a pasture or of staying aloft while matching the speed of a slow-moving automobile. If the U.S. military possesses an aircraft with these miraculous capabilities, it would seem useful to unleash it against our acknowledged enemies. A totally silent black triangle that can hover, float, search, zip away at incredible speeds, and disappear at will would come in handy, no?

Similar mystery triangles have been seen with increasing frequency in recent years all over the world. A 2004 analysis by NIDS scientists and published on the institute's website examined more than five hundred cases from across the United States, many of which contain details similar to the descriptions offered by the Gormans. The NIDS analysis of these incidents strongly suggests that while some of these craft could represent an unacknowledged deployment of classified military technology, or perhaps a behind-the-scenes, legally questionable surveillance effort over U.S. soil, a growing body of information points to other possible conclusions. It is conceivable that some mystery triangles, including those seen over the Gorman ranch, may belong to someone other than the U.S. military and are truly unidentified.

While similarities exist among nearly all of the UFO events that occurred on or near the ranch and other well-documented cases from other locations around the world, there is one phenomenon for which there are few, if any, known precedents. Tom Gorman's descriptions of the blue orbs filled with swirling liquid, electrically charged orbs that seemed to be under intelligent control, may be unique in the annals of UFO and paranormal research. And as such, these orbs may represent the best clue to what was really happening on the Gorman ranch.

CHAPTER 19

The Tunnel

Augost 25, 1997. The night was warm, clear, and beautiful. The NIDS team was sitting silently on the edge of a bluff and the scene one hundred feet below them was idyllic. Trees in full leaf edged the small pasture, and in the distance, two fields away, the ranch animals grazed peacefully. The scattered yard lights of the neighboring homestead punctuated the darkness. In the distance, some coyotes howled. Investigators Jim and Mike (not their real names) were on the bluff because it afforded them the best view to monitor the area where, in the previous few months, several strange events had taken place.

Two colleagues were about a mile away on the far side of the property monitoring another hot spot. Mike and Jim had been in sporadic contact with them every couple of hours, but the agreement was to stay out of radio contact unless something really eventful began. But nothing stirred.

Some four hours into the watch, Jim, the most experienced investigator, had quietly climbed down into the small pasture and had sat in the middle of the field to meditate. He had found over the years that meditation sometimes activated the "phenomenon," whatever that was, although it happened too few times to be anything more than an anecdotal observation. Nothing happened.

At 2:30 A.M., after a six-hour watch, they decided to move their operation to a different part of the ranch. Quickly, they began to disassemble the camera from its tripod and pack up the two equipment carriers filled with the tools of the scientific investigator's trade: cameras, portable magnetic field detectors, and night-vision binoculars. In 1997, their generation-three binoculars were considered state of the art. They worked by amplifying low-level ambient light through a series of hi-tech photomultiplier tubes. They were far superior to the usual Russian night-vision equipment that looked just in the infrared. Through these night-vision binoculars, the scene appeared to be in daylight with sharp crisp outlines, not the fuzzy, wavy images visible through the Russian equipment.

Just as Jim was thinking that his meditation hadn't produced a damn thing, his eye caught a very faint light on the track 150 feet below him. He watched it, mildly puzzled, thinking that it might be a small piece of glass on the track that was reflecting ambient light. It was a faint yellowish color, and as he watched, it appeared to be growing brighter. Twenty seconds later, he nudged Mike. It was definitely getting brighter and, as both of them watched, it seemed slowly to be getting bigger.

"Hand me the camera," Jim muttered. At the same time, Mike quietly and efficiently unpacked the night-vision binoculars he had just put away. Jim set up the tripod and positioned his camera loaded with infrared film in line with the light that had now grown to six inches in diameter. It was still a dull yellow but had definitely grown brighter. Carefully, Jim set the shutter to thirty seconds, reasoning that a long exposure might capture this mystery light on the freshly loaded roll of infrared film. He was ready to pop off all thirty-six shots if necessary.

As Mike brought the binoculars to his eyes, Jim heard the sharp sucking in of breath. Jim could see that the light was now more than a foot wide and was still growing larger. This very obviously was not a reflection. The dirty yellow expanding light seemed to be positioned just above the ground, rather than directly on the ground, but Jim could not be sure. "It's a tunnel, not just a light," Mike whispered. Jim ignored his partner's growing agitation as he increased the length of the time exposures to forty, then fifty seconds. Mike was now standing up. "Jesus Christ," Mike said hoarsely. "Something's in the tunnel!"

Jim looked carefully at the light below. It had now expanded to more than two feet. Something that big should definitely register on his film. "Oh, my God," Mike said suddenly, thoroughly frightened. "There is a black creature climbing out. I see his head." Jim felt alarm. His companion was plainly bordering on panic.

"It has no face," whispered Mike. "Oh, my God, it just climbed out." Jim rubbed his eyes and shook his head. All he could see was a dirty yellow light, now about four feet in diameter, a hundred feet below him. Why couldn't he see what Mike was seeing? Suddenly, it dawned on him. The binoculars. He motioned for Mike to hand them to him, but not before increasing the time exposure to ninety seconds.

Mike ignored him. "It's on the ground," he said. "Oh, my God, it walked away." As Mike danced on the ledge a few feet away, plainly in a panicky state, Jim could see the light decreasing in size. Within thirty seconds, the dull yellow circle had shrunk to about half its full diameter and was losing intensity. In the meantime, Jim pulled Mike over close to him and asked, "What happened?"

"A big black creature just crawled through that tunnel, got onto the ground, and walked away," Mike said. "That's what happened. And it's lurking around here somewhere."

Jim felt a chill. "I only saw that yellow light," he said doubtfully. "Are you sure?"

"Jesus Christ, of course I'm sure," Mike replied. "The night vision turned the light into a three-D tunnel, and a large creature, I am thinking maybe four-hundred pounds, at least six feet tall, just crawled out of the damn tunnel." Mike was sweating profusely and still breathing quickly, but he seemed to be regaining control over himself. Jim looked carefully around and motioned Mike to be quiet. Carefully, they listened for any noise of displaced stones that might indicate something climbing up the cliff in the darkness. By that time, the yellow light had gradually faded and was no longer visible. Only a deep silence remained. Nothing seemed to move. Even the distant coyotes had stopped howling.

After fifteen minutes or so, they grabbed some detectors and clambered as quietly as possible down the steep, rock-filled gradient below them. It would be very easy to break an ankle coming down in the darkness, so they took it very slowly, stopping every few feet to listen for sounds of the creature. Nothing stirred. But once near the track, a strong, pungent odor assailed their nostrils. They were both familiar with this sulfur-laden odor. It seemed to be centered on the spot where the light or tunnel had been.

Jim felt slightly nauseous from the pungent smell. Quickly, Mike scanned the ground in a twenty-foot-diameter circle for any signs of radiation. The Nardalert counter could pick up alpha, beta, gamma, and X-rays, but after minutes of careful monitoring, only background levels were detectable. Meanwhile, Jim slowly scanned the area with a Trifield meter to detect any unusual magnetic spikes. Nothing registered.

As the time passed, their confidence slowly returned. It had been almost thirty minutes since the creature had vanished, and nothing had come rushing out of the night to attack them. They were still watchful as they finished the monitoring and then climbed back up to the top of the bluff to retrieve their equipment. Again they scanned the area with night-vision equipment for anything out of the ordinary, and slowly they began the half-hour walk back to the observation trailer.

Mike described to Jim what he had seen, the bizarre and creepy sight of a huge black humanoid form using its elbows to lever itself along a three-D tunnel that appeared to be suspended a couple of feet above the track on which they now stood. Jim believed him. He knew that Mike had spent long hours in the darkness tracking this stuff and was not given to flights of fancy—at least not until tonight.

The team spent hours the following day searching for footprints. But the ground was hard and none were visible. Mike and Jim worked on a written report of the incident. Their instruments had failed to record anything unusual. The photos were disappointing, showing only a single very faint blurry light in one and nothing on the rest of the roll of film. What had Mike witnessed crawling through the strange three-dimensional tunnel? What was that large, featureless, bulky humanoid creature that stood up and walked away silently into the darkness?

CHAPTER 20
Monsters

Human beings have been terrorized, mystified, and fascinated with tales of mystery beings and unearthly creatures for as long as our species has existed. Monsters, demons, goblins, and ghosts have their roots in oral traditions that date back to cave-dwelling yarn spinners. Over the centuries, there have been innumerable other entries in this pantheon of the paranormal, from witches and werewolves to vampires and sea serpents, from blood-sucking ghouls to blood-sucking *chupacabras,* from trolls and leprechauns to bug-eyed space aliens who arrive in the night to snatch unsuspecting people from their beds. Godzilla, Freddie Krueger, and Frankenstein's monster didn't exist, so they had to be invented, the product of fertile imaginations and cinematic special effects artists. If they weren't real before, they are now, imprinted forever on our collective psyche.

There is a distinct difference between monsters that exist only on celluloid or the printed page, however, and those that occasionally make overt intrusions into our personal realities;

one emerges from the supernatural, while the other, like Bigfoot, has distinct roots in our flesh-and-blood reality. Although Mike is certain that the creature he saw crawling through the tunnel lacked hair, there is no question that Gorman and members of his family witnessed some Bigfoot-like creatures on the ranch on more than one occasion, as have others in the area.

Contrary to popular impression, however, Bigfoot wasn't born in the forests of the Pacific Northwest. He didn't just spring up from nowhere alongside Starbucks Coffee, Microsoft, and the grunge rock of Nirvana. Long before Europeans arrived in North America, indigenous peoples throughout the continent knew of Sasquatch, which is the name used by the Salish tribe of the Northwest. Coastal tribes called him Bukwas or Dzunukwa and have featured his likeness in masks and on totems dating back to the 1700s. The Lakota Sioux of the northern plains referred to Bigfoot as Chiye tanka, roughly translated as the big elder brother. Joe Flying By, a Hunkpapa Lakota, told Peter Matthiessen, the author of *Thunderheart*, that "the Big Man is a husband of the earth," a being from ancient times who can take a hairy form or even change into a coyote.

"There is your Big Man standing there, ever waiting, ever present, like the coming of a new day," Joe Flying By was quoted as saying in an article in *The Track Record.* "He is both spirit and real being, but he can also glide through the forest, like a moose with big antlers, as if the trees weren't there."

Canadian writer and Bigfoot researcher Ron Murdock asserts that the existence of Sasquatch is taken for granted throughout Native North America. The Athabaskan Indians of Alaska regard Sasquatch as a "big brother" who looks out for Native American peoples and brings signs or messages during troubled times. Hopi elders believe that Bigfoot delivers messages from the Creator, often messages about man's disrespect for harmony and balance. "In Native culture, the entire natural world is seen as a family. Sasquatch is regarded as one of our closest relatives. In each tribal dialect, there is a word for Sasquatch," Murdock writes.

When white trappers, hunters, and explorers first made contact with native tribes in North America, they often heard the stories about a race of large, hairy humanoids. A few of these mountain men had their own encounters, some of which were reported in frontier newspapers of the day. In 1784, the London *Times* reported the capture of a "huge, manlike, hair-covered creature" in Lake of the Woods, Manitoba. In 1811, a fur trader named David Thompson told a journalist about his sighting of a "hairy giant" near what is now Jasper, Alberta. In 1840, a missionary named Elkanah Walker, who spent nine years living among the Spokane Indians, described the tribe's belief in the "existence of a race of giants . . . which hunt and do all of their work by night" and leave behind a "track about a foot and a half long." The *Butte Record* of northern California reported a sighting of a "male gorilla or wildman" in November 1870. The *Victoria Colonist* of British Columbia reported the capture of a "gorilla-type creature" in July 1884.

Outdoorsman and future president Theodore Roosevelt reported his own Bigfoot story in his 1893 book *Wilderness Hunter*. Roosevelt met a trapper named Bauman, who told of a deadly encounter with a tall beast that invaded a campsite in British Columbia, smashed all of the gear in the camp, and haunted the vicinity with a sound described as a "long-drawn moan" that struck Bauman and his partner as "peculiarly sinister." Roosevelt wrote that the beast killed Bauman's partner while Bauman was away from the camp and that "the footprints of the unknown beast-creature, printed deep in the soft soil, told the whole story." The mystery animal left four deep fang marks in the neck of the victim but did not eat the body. After rolling around in the dirt, the attacker "fled back into the soundless depths of the woods."

It is possible that at least some of these accounts represent mistaken identifications of bears or other large animals, except for the fact that the witnesses were all experienced hunters and lifelong outdoorsmen, people who were intimately familiar with the behavior, sounds, and physical features of bears and other species known to inhabit the forests and mountains. The Native American tribes that have lived and hunted for centuries in these same areas would presumably know the difference between a bear that stands on its hind legs and a "hairy man." What's more, as the nineteenth century gave way to the twentieth, and the western wilderness became more populated, the sightings and encounters continued.

In 1924, a hunter named Albert Ostman had a dramatic encounter with what he said was a family of Bigfoot beings while on an outing near Vancouver Island. In the same year, the Portland *Oregonian* reported the story of Fred Beck's hunting party on the slopes of Washington's Mount Saint Helens. Beck and his companions claim they used their rifles to fend off an attack by a pack of rock throwing "mountain gorillas," which they also described as mountain devils. A teacher named J. W. Burns, who spent many years living among the Chehalis tribe near Vancouver, wrote several regional newspaper articles in the late 1920s about encounters between his Indian friends and a race of hairy giants. In 1927, one of Burns's accounts was printed in a national magazine but didn't seem to make much of a lasting impression on the larger public. However, the Burns articles did introduce to readers a name for the beast. The name was Sasquatch.

The Sasquatch story received only sporadic coverage in local newspapers around the country until 1958, when a road crew working in a remote area of northern California made a plaster cast of one of the huge, humanlike footprints that had been found in the mud around their work site over a period of several weeks. The site foreman took the cast to a local newspaper. The story was then picked up by wire services, and before long, the whole country was abuzz about "Bigfoot."

Although there have been few, if any, legitimate scientific in-
quiries into the Sasquatch legend, scientists decided long ago
that the story was absurd on its face, that it is most likely the result of hoaxes or mass hallucinations, or has been promulgated
by media coverage. But since the story has been around since the
1700s, and has thrived among indigenous tribes and remote populations, the explanation that it was created by mass media
seems weak. Certainly there have been some attempted hoaxes
over the years, but no single hoaxer or group of hoaxers could
possibly account for all of the sightings over so much time and
so much territory. And mass hallucinations do not easily cross
from one culture to another. The Bigfoot story is not only multi-
cultural, but also international in scope.

What's more, there is a body of evidence that suggests this
mystery is worthy of serious inquiry. After all, the confirmation
of the existence of a previously unknown primate species, especially a species whose adults are commonly described as being
seven feet tall or taller, one that has been living for centuries
under our noses in the forests of North America, would be one
hell of a scientific news story.

Thousands of eyewitness accounts have been recorded over
the past three centuries. One Bigfoot, in particular, seems to
have been spotted by three different groups of witnesses. In 1988,
a father and son spotted what they described as a six-foot
Sasquatch near a creek in Grays Harbor County, Washington.
While most Sasquatch are said to be covered with dark hair, this
one was white, with blue eyes and a pink complexion. The father
and son were within twenty feet of the beast, and they noticed
that it seemed to move with a limp. After the encounter, the two
examined the large tracks that had been left behind and determined that the animal's right foot was crippled.

Then in July 1995, another father and son were on a camping
trip at a reservoir in the same county. While on a hike overlook-
ing the water, they saw two Sasquatch below. The beasts ap-

peared to be playfully wrestling with each other. One was
brown, the other white. The white one walked with a limp. The
sighting lasted more than twenty minutes, and before the two
Sasquatch walked off into the forest, six other hikers joined the
father and son as they watched the creatures cavorting.

Finally, in October 1996, a husband and wife were on a log-
ging road in the woods of the Oregon Cascades. The husband was
using a chain saw to cut up firewood while his wife sat in their
truck. Two other men were cutting wood about one hundred
yards away. The witnesses were startled when a Sasquatch "as
white as a towel" emerged from the trees some twenty-five feet
away from the two men, then walked back into the forest with a
pronounced limp. The husband, Frank, didn't see the Sasquatch
but was told by the other two men that it had blue eyes and a
pink complexion. When Frank went looking for any telltale
tracks, he found them and later made plaster casts of two large
prints, one from each foot. The cast of the right foot revealed
that the Sasquatch was missing his large toe, which would explain the limp.

What are the chances that three groups of multiple witnesses
in different years and different places would see a white, pink-
faced, blue-eyed Sasquatch that walks with a limp? Slim, it would
seem.

Private Bigfoot research organizations have accumulated huge
volumes of sighting reports. Other types of physical evidence accompany some of the sightings. For example, in Alpine, California, a Sasquatch was seen eating apples from a tree in front of
the house. Not only did it leave behind monstrous footprints,
but the tree and a few others were found to have been stripped
of fruit in their upper branches, at a height not reachable by a
man. In Arizona, a mother and daughter watched as a seven-foot
Sasquatch looted their garden of corn and turnips, and left behind characteristically large footprints and a decimated patch of
ground.

Sightings of hairy ape-men have been recorded in forty-nine of the fifty states and in every Canadian province, but the phenomenon is by no means unique to North America, or to modern times. Ancient Mesopotamians wrote of Enkidu, a wild man who could communicate with animals. The Bible mentions Esau, whose body was supposedly covered with animal hair and who emanated a strong, offensive odor. Early European writers recorded stories of "the wild men of the woods," seen in Germany, France, and other countries. Hairy ape-men have been reported in China, Nepal, Russia, Ireland, Thailand, Vietnam, India, Malaysia, Indonesia, New Guinea, New Zealand, Australia, Brazil, Argentina, Colombia, Panama, Guatemala, Kenya, and other countries. In Australia, the wild men are known as Yowies. In the Himalayas, of course, the more famous name is Yeti, also known as the Abominable Snowman, which has been seen by mountain peoples for centuries. Tenzing Norgay, the famed Sherpa mountaineer who guided Edmund Hillary to the summit of Mount Everest in 1953, told writer John Keel many stories about Yeti encounters, as did the lamas, trackers, and villagers of the region.

"To dismiss all of this as collective hallucination, the primitive need for the mythological, or simply as archetypal legends common to all of mankind," writes anthropologist Helmut Loofs-Wissowa in the *ANU Reporter,* "will not do anymore."

Other than the eyewitness sightings, the footprints left by these animals represent the most direct physical evidence of their existence. More than seven hundred plaster casts of Bigfoot prints have been collected over the years. The average length of the feet is 15.6 inches. The average width is 7.2 inches. (The foot of a 7'3" basketball player might be 16 inches long but only 5.5 inches wide.) Needless to say, these are very large feet. Clearly, something is leaving these prints in the deep forests. It boils down to either hoaxers or an unknown species. If hoaxers are responsible, they are incredibly diligent, since the tracks have been found in remote areas all over the world for many, many years.

Have some of the collected prints been faked? Without question. But what about the rest of them? Oregon Regional Primate Center's W. Henner Fahrenbach conducted a study of more than five-hundred footprints collected over thirty-eight years from a wide swath of North America. Fahrenbach notes that many of the footprint records were obtained by tracking the animals for miles, not for a step or two. Since most of the prints in the database were collected by individuals who don't know each other and don't know what the "normal" size of a footprint should be, Fahrenbach thinks it is unlikely that a few cases submitted by hoaxers could have much of an impact on the overall validity of the database. Using a Gaussian distribution of the footprints and the locations where they were found, Fahrenbach says the patterns are similar to those of known animals. In other words, the prints indicate that this is a real species of unknown animal that is living out there in the wilderness and is not the work of a few hoaxers.

A latent-fingerprint expert named Jimmy Chilcutt, who works full time for the Conroe Police Department in Texas and who is highly regarded by the FBI and DEA for his expertise, decided to enter the Bigfoot debate a few years ago. His intention was to debunk the footprints that had been collected. Although he did find some prints that he believes to be fakes, Chilcutt's views about Bigfoot changed the more he dug into the evidence, and he now believes that this unknown animal species really exists.

Hair samples that have been collected in areas frequented by these creatures represent another intriguing form of evidence. Hair samples obtained in the Blue Mountains of Washington State have been analyzed by the Department of Molecular Genetics at Ohio State University. The DNA that was extracted was too fragmented to allow meaningful gene sequencing. Some of the hair samples were identified as being from known animals. A few were shown to be synthetic. But others are believed to be from a "non-human but unknown primate," although, at best, the test results are still considered inconclusive.

In 1988, a team of one hundred Chinese researchers associated with the China Wildman Research Center converged on a mountainous Hubei province where sightings of a mysterious man-ape date back almost three thousand years. The researchers found and analyzed numerous hair samples. They identified several hairs that do not belong to any known species in the area and concluded that the hairs prove the existence of a "rare and advanced primate that is similar to man."

British scientists have supposedly analyzed hair samples found in the kingdom of Bhutan, where the locals refer to their own version of hairy biped as the Migyur, and concluded that the DNA isn't from a human or a bear or any other known species. For the most part, however, mainstream scientists do not find the evidence compelling, nor have they been impressed with audio recordings of alleged Bigfoot cries or by the few film snippets that have largely been dismissed as staged fabrications. Scientists rightly ask where are the skeletons of dead Bigfoot. Why is there no fossil record of these creatures as they evolved through the ages? Bigfoot proponents note that there isn't much of a fossil record for chimpanzees or gorillas either, yet we know both species exist. They also counter that no skeletal remains of gorillas were found until long after the species itself had been discovered. The pro-Bigfoot crowd believes, with some justification, that modern science hasn't fairly evaluated the evidence and testimony.

There is one other possible explanation for the lack of Bigfoot bones and a fossil record: What if the creatures are not flesh-and-blood but are of a paranormal nature? There are, in fact, scores of cases where witnesses have seen Bigfoot-type creatures in a paranormal context. There are cases in Australia and the United States where families have been plagued by what seems like prototypical poltergeist activity, and then they discover a Bigfoot creature on their property. Bigfoot animals have also been seen in the vicinity of animal mutilation cases. In 1974, when a wave of cattle mutilations occurred in Nebraska, Kansas, and Iowa, witnesses reported seeing strange lights in the sky and animals that resembled apes or bears on ranches where mutilations occurred. In Boise, Idaho, three separate witnesses saw hairy, manlike beasts in their yards in July 1975 during a wave of mutilations and UFO sightings.

The apparent paranormal aspects of many Bigfoot accounts make many cryptozoologists uncomfortable. In fact, most Bigfoot researchers choose to downplay some of the more exotic aspects of their research. After all, they believe that Bigfoot, Yeti, and their assorted cousins are flesh-and-blood beings that just haven't been found yet. Whenever witnesses discuss Bigfoot in connection with such topics as UFO sightings or psychic phenomena, cryptozoologists get understandably antsy. After all, mixing in a plethora of other weird topics could not only strengthen the hand of Sasquatch skeptics but also give scientists and journalists another reason to sidestep any serious examination of the Bigfoot mystery.

However, it simply isn't honest to discard pieces of information that might not fit a preconceived idea of "the truth." Tony Healy, who has devoted twenty-five years of his life to pursuing the Australian Yowie, estimates that to exclude all Yowie stories that smack of the paranormal would mean the elimination of about 20 percent of the case files. Healy and other researchers suspect that these creatures have some sort of sixth sense that assists them in avoiding detection or capture. Healy says a hefty percentage of Yowie sightings involve the presence of UFOs, other unknown animals, even lake monsters. Healy, for one, has now concluded that the Yowie is some sort of shape-shifter, a phantom that may always remain beyond human comprehension.

Many Native American tribes have long regarded Sasquatch as something more than a flesh-and-blood animal. Ray Owen, whose father was a prominent spiritual leader on the Prairie Island Reservation, offers the opinion that the Sasquatch "exist in another dimension from us, but can appear in this dimension whenever they have a reason to."

Owen explains, "It's like there are many levels, many dimensions. When our time in this one is finished, we move on to the next. But the Big Man can go between. The Big Man comes from God. He's our big brother, kind of looks out for us."

Joe Flying By has made a similar observation. "I think the Big Man is a kind of husband of Unk-ksa, the earth, who is wise in the way of anything with its own natural wisdom," he says. "Sometimes we say that this One is kind of a reptile from ancient times who can take a big hairy form. I think he can also change into a coyote."

Certainly there are plenty of reports in the literature to bolster the idea that Bigfoot might be even stranger than he seems. One oft-repeated characteristic of Bigfoot encounters is the feeling of "nameless dread," a pervasive, bone-chilling fear that seems entirely out of proportion for the nature of the encounter. Witnesses, even those with a healthy curiosity about the unknown, have told of being gripped by uncontrollable fear that causes some to faint. Even fierce hunting dogs have been known to cower and whine in the vicinity of these creatures. Some witnesses are overwhelmed with fear even before they see the Bigfoot.

Witnesses on the Gorman ranch know this feeling of dread well.

CHAPTER 21

Dulce

By the end of the summer 1997, the NIDS team had spent about a year hunting a very elusive prey. The initial promise of obtaining data and personal sightings of phenomena had slipped into a strange cacophony of bizarre, unrepeated, transient encounters with something that was difficult to explain. Certainly nothing that had happened so far could be called scientific data on aerial phenomena, which was NIDS' original mission. The plan for the Utah ranch had been to quickly amass conclusive evidence of a sustained UFO event with video, camera, and other instruments. But whatever had been playing cat-and-mouse games with NIDS personnel was simply too fleeting for us to capture a spectrum of a light or a good video sequence that could be productively subjected to analysis. In short, the abrupt, unexpected events seemed almost designed to evade capture.

As the team continued hunting the phenomenon during 1997, we decided also to begin a second front on another hot spot, with the hopes of gathering enough data from a second location to provide us with a comparative overlap. The area around Dulce, New Mexico, became the focus of my attention for a good part of 1998 and 1999.

The remote town of Dulce lies along Route 64 and has remarkable parallels with the Skinwalker Ranch and the Uinta Basin. Dulce is an isolated town of maybe two thousand people, largely populated by Native Americans, just like the ranch. The Jicarilla Apache reservation completely encompasses Dulce. Also, like the ranch, it is hidden away from the main interstate highway system. It takes an effort to get there.

Dulce lies in a valley surrounded by the majestic peaks of mountains. When you enter the town you immediately sense how run-down the place is. The main street is windswept, and hungry stray dogs search for sustenance. There are no movie theaters and no shopping mall, just a large Best Western hotel that is undoubtedly the town jewel. And near the hotel is a liquor store. The town's prime entertainment is obvious. Victims of alcoholism can be found lying against street corners, a testimony to the hard life. Until a couple of years ago the local gambling casino was the center of the town's life, but it closed amid charges of widespread looting of the coffers by unnamed town elders.

At the same time, Dulce has earned a worldwide reputation as a center for paranormal activity. Like the Uinta Basin, in the past thirty years this small, impoverished town has been host to a bewildering assortment of cattle mutilations, UFO activity, Sasquatch sightings, and numerous other oddities. It is accurate to say that Dulce and its surrounds can easily rival the Uinta Basin in northeastern Utah as a hot spot of anomalous activity.

The Dulce area became famous in the late 1970s when a young New Mexico state patrol officer named Gabe Valdez began to investigate, and later report, a series of bizarre mutilations that occurred to the cattle owned by the Gomez family. The Gomez property lies just a few miles outside the center of town.

Valdez was an aggressive investigator and he quickly established a reputation in the Apache community as somebody who did not back down in the face of danger. In the numerous bar fights, homicide investigations, and spousal abuse incidents that characterized everyday lawlessness in Dulce, Valdez was often on his own and frequently outgunned. But he never backed down. When the mutilations began to happen on Manuel Gomez's property, Valdez was right there.

Gomez had begun to complain that someone or something was killing his cattle. The animals were all clustered in a large pasture within a couple of hundred yards of the household. The cattle were usually found the next morning with their eyes, ears, sometimes tongue, rectal region, and reproductive organs removed. Valdez quickly realized that these were not predator or scavenger attacks. The cuts on the animals were precise and usually bloodless. There was definite skill, even artistry, involved. For people who had worked most of their lives around cattle, it was easy to distinguish between the jagged edges and messy blood and guts strewn through the area of predator and scavenger attacks versus the clean, usually bloodless cuts in cattle mutilations. There was no contest. Manuel Gomez eventually lost more than twenty head of cattle and basically went out of business.

Valdez also began tracking strange, round, orange-colored objects that flew around the area and were sometimes observed where a dead cow would later be discovered. Sometimes these objects were the size and shape of a harvest full moon, but unlike the moon, these objects moved in jerky, erratic paths across the sky. Many local Dulce people reported seeing the same objects during the mid- to late 1970s.

I first met Valdez in 1997 when he came to work for the National Institute for Discovery Science. As a parallel project with the Gorman ranch, he and I spent about one hundred days in Dulce. The object of the exercise was to investigate the alleged

anomalies that had made Dulce legendary. Valdez had by then retired from law enforcement and was only too happy to continue
the investigations with NIDS. After my first trip to Dulce, I realized why most of the amazing events that took place in Dulce
had never been publicized: nobody had ever cracked the *omertà*,
or the code of silence, of the Jicarilla Apache tribe.

Valdez was treated like royalty here. On my scores of visits
with the stocky, genial police officer, people would run up to
him to shake his hand. They would honk their horns as he
drove by. It was like accompanying the pope to Dulce. I often
suggested to Valdez that he should run for mayor of the town.
He would win in a landslide. Valdez would just grin at my suggestion. And of course, the normally reticent inhabitants of
Dulce were only too eager to open up to somebody whom
Valdez recommended. Over the span of two years, Valdez and I
conducted more than seventy interviews with Dulce residents
and cataloged a stunning variety of anomalies that had never
before seen the light of day.

A study of activity in New Mexico was also useful to NIDS
because, like northeastern Utah, Dulce was a desolate area peo-
pled mostly by low-income folks and a large proportion of Native Americans. The area, like northeastern Utah, was far away
from the main interstate highway system. Some of the incidents
we gathered information about while we were in New Mexico
bore no resemblance to those seen on the Utah ranch, while
others were similar. What was striking, however, in both New
Mexico and in Utah was the sheer variety of the objects seen,
and events experienced, by scores of people.

In the late summer of 1979, Terrance Tafoya and Charlie
Mundez (not their real names) were relaxing after working on
the road by Mount Archuleta just outside Dulce. They were sitting on the tailgate of their truck. It was 8:30 at night and the
two were enjoying the peace. Suddenly, over the treetops, about
two hundred feet away, a silent silver disc appeared. It was about

150 feet in diameter, metallic, and had a dome on top. It was moving slowly, smoothly, and absolutely quietly. They could easily make out a series of metal struts underneath the craft as it passed over them. Two sets of lights, one blue and one yellow, rotated slowly in a counterclockwise direction underneath the giant craft. The two watched dumbfounded, as the object hovered no more than 150–200 feet above them for about five minutes. They were amazed that such an object could even fly. The object then tilted at an angle of about forty-five degrees and began climbing slowly. Then from a stationary position it suddenly and noiselessly took off across the valley. The eyewitnesses estimated that it took about four seconds to cover the sixty miles over to the distant peaks where it disappeared from their view. The object left a flash of light in its wake.

Although the object was bigger than the one seen by the Gormans in September 1996, the silver disc bore resemblance in shape to the object seen in Utah. In short, it was a classic "UFO." Partly because of the multiple incidents reported by people around the Mount Archuleta area, NIDS began a surveillance program with personnel and equipment on the top of Mount Archuleta. The program was nowhere near as intensive as the one carried out in Utah, but such was the breadth and scope of the previously untold reports that NIDS began to monitor the area.

We tracked down and interviewed both Charlie Mundez and Terrance Tafoya separately. Dulce interviews required a special technique. We learned to arrive unannounced in the village and simply to drive around the town for a couple of days asking quietly about the whereabouts of our target witnesses. Over the next couple of hours we would gradually track them down and then walk up and nonchalantly begin a conversation. Dulce at that time did not have a great number of telephones that actually worked, so making telephone calls in an attempt to track down witnesses was a waste of time. And once again, Gabe Valdez was invaluable because he not only knew the witnesses but where to find them.

Charlie Mundez, then in his late sixties, told us he had also seen something else. He was fifteen years of age at the time and was driving in the direction of Dulce with his family. It was still daylight at 5:30 on a summer's day. When they crested a rise, the family all saw a huge craft hovering above the valley. Mundez's recollection was that the craft was shaped somewhat like the Tacoma dome; it had a flat bottom and it stretched from Dulce Lake all the way across to Archuleta Mesa. Mundez declared that the object was probably five miles across. He described a single horizontal row of windows that spanned the middle of the giant craft as it hovered peacefully over the valley. He also said that he could vaguely make out the landscape behind the object, making it slightly transparent. Yet he could also easily see the boundaries of the giant object. The family watched in silence as they continued to drive toward their home in Dulce, but when they arrived and got out of their vehicle and looked up, the object was gone.

We heard of many similar incidents from witnesses around Dulce, many of whom had never told anybody outside the tribe before, with some of the incidents going back to the early 1950s, just as they did in the Uinta Basin. This may be unsurprising given the rash of sightings of UFOs nationwide that occurred at the time, but we felt it was significant that so many people in such out-of-the-way places as Utah and northern New Mexico had witnessed such a wide variety of craft and creatures at around the same time.

We also separately interviewed two very high-ranking officials from the Jicarilla Apache tribe who told us of an incident that had happened in January 1996. Nine people witnessed this incident as they traveled in four cars in a convoy at about 11:30 on a clear moonlit night on the way back from a basketball game. Visibility was superb. One of the eyewitnesses said that he saw something way off in the distance that looked like a set of pylons with blue and red lights on top, somewhat like a large natural gas platform there. The sight confused him because he knew the road well and had never before seen a natural gas platform. Because of their official positions and involvement in local politics, both have requested anonymity.

As the cars drew nearer, they saw that the "natural gas platform" was actually a huge flying object slowly moving toward them. It looked as if it was just about to crash into the cliff on the edge of the canyon. One of the officials reached for his cell phone in order to report an aircraft accident and to alert search-and-rescue teams when he noticed that the object was now moving slowly directly over the four cars as they drove through the canyon. The object was huge, one hundred to two hundred feet above them. According to the second official, the object was so big that it spanned the entire canyon from wall to wall and he still could not see the ends of it. This would make the craft more than a mile across.

As it moved slowly over their cars, both people said the underside consisted of rows of dull metal sheets overlapping like roof slates. The second official described seeing large rivets about a foot to two feet in diameter and described the shape of the object from underneath as roughly circular. The first official said that he did not know the shape from underneath but that the object from the side had three large domes on top. The second official pulled his car over and got out just in time to see the aircraft disappearing over the canyon wall. There was no sound. The object seemed instantaneously to appear a long distance away as if it had silently moved at great speed.

Following the interviews, I drove with Valdez to the precise spot in the canyon described by the two witnesses. Indeed, the approximate distance from canyon wall to wall was just under a mile, and for a single object to have spanned the entire canyon, flying above it, implied an almost impossibly large craft. Both eyewitnesses in separate interviews admitted being extremely puzzled as to how an object of such dimensions could move through the air so smoothly, so silently, and so easily. The kind of huge object that flew over Cordoba Canyon would later become famous around the turn of the millennium as thousands of people began reporting football-field-sized objects flying low over populated areas.

At 3:30 on the afternoon of November 17, 1984, Bruce Montoya and a friend drove onto another friend's property located about seven miles north of Dulce, just over the Colorado border. It was a cloudy afternoon and slightly muggy. They dismounted from their truck to sit on the porch. Bruce's AK-47 was leaning against the wall of the cabin. Their dog, which was tied to the porch, began barking. They followed the dog's gaze. Coming slowly across the pasture, about fifty feet up in the air, was the strangest thing they had ever seen. The nearest Bruce could describe it to me was that it looked like a gray manta ray. It was no more than fifty yards away at the closest point to them. The object had upraised wing tips and was about one hundred feet in diameter, but without the tail that a stingray has. It moved at about five miles per hour across their line of sight. The dog became very silent as the object made a noticeable *whoof-whoof* sound. The two witnesses watched in amazement as the object moved past them, then slowly tilted so that they could see three perfectly circular portholelike structures underneath.

Bruce reached for his AK-47, rapidly inserted a clip into the breech, and looked through the telescopic sight. He focused on the surface of the craft as it moved slowly in front of him. Without knowing what he was doing, he slowly squeezed the trigger. The gun jammed. The craft continued to move slowly in front of him, beginning to angle away toward the treetops a couple of hundred yards away. The slow movement seemed almost dreamlike. After a couple of minutes, the craft drifted over the treetops and seemed to become slightly transparent. Then there was a flash, and although he couldn't be sure, he thought that the object had moved away from them at an impossible speed.

Montoya later speculated that the object might have been a creature. Its rough, leathery skin with ridges and dimples reminded him strongly of whale skin or rough sharkskin. It seemed and "felt" biological to him. Was this a creature or a craft? The manta ray shape was vaguely reminiscent of the black Stealth fighter–like object that Gorman saw on the snow-covered ranch in the winter of 1995–96. The way the black object reacted to Gorman's slight noise, swiveling suddenly toward him and extinguishing all lights, was much like a wild beast would react. And did the object or creature speed away or disappear into another dimension, as Gorman often wondered of the strange sights on his property?

From the window of his house, Montoya has repeatedly seen small bouncing lights zigzagging over the ridge where several well-known Dulce mutilations occurred in the 1970s and the 1980s. He told us of looking out his widow at about 2:30 one morning in October 1998 and seeing two whitish balls of light flying alternately apart and then together. The balls moved very quickly up and down, zigzagging, and executing right-angled turns. Montoya said that this has been a common sight in Dulce for years.

Dulce in the 1960s and the 1970s was also home to unusual creatures, just as the Utah ranch was. In 1962, Sheila Bromberg was returning home late after working at a local restaurant. She thinks it was about 1 A.M. when the strange event happened. Sheila is a small, quiet woman close to seventy years of age, with bright eyes and an engaging manner. She has spent the last twenty years of her life helping out at the tribal center in Dulce. In her quiet way she is passionate about the plight of the Apache tribe in Dulce, the devastating effects of alcohol, chronic disease, lack of education, and other social problems. She does what she can. She was very reluctant to talk with us but eventu-

ally agreed to meet with us at the Jicarilla Apache social center—
the Best Western Hotel, in other words. We bought her lunch
and spoke casually.

Sheila told us she was glad to see her cat standing to greet her
that morning as she walked to her doorway. When she lifted the
cat, the animal suddenly arched its back, hissed loudly, and dug
its nails into her shoulder. This was highly unusual behavior for
her pet. The cat was focused on something behind her, but when
Sheila looked over her left shoulder, she saw nothing. Then, with
the cat still hissing and plainly alarmed, Sheila looked over her
right shoulder and saw a small figure standing nearby. She nearly
jumped out of her skin.

It was not human. It had a large head, large eyes, and grayish
wrinkled skin. Sheila nearly screamed. The creature simply stood
there, unmoving. Sheila quickly opened the door of her home
and with the cat still on her shoulder ran inside, where she
stayed for some time. When she plucked up the courage to go
outside, she circled her house more than once, but the creature
was nowhere to be found.

It is worth noting that Sheila's experience happened in 1962,
about twenty years before the huge media publicity regarding
the "Grays" and nearly thirty years before the publication of
Whitley Streiber's famous book *Communion* with its unforget-
table painting of the creature on the cover. Sheila's experience
has the ring of authenticity because it predated all of that publicity and also because she had told only her closest friends in
Dulce about it. We confirmed that she had actually described
this event to others back in the early 1960s.

In the 1970s, Sheila saw a large Sasquatch-like creature standing less than a hundred yards away from her. The creature was
over seven feet tall. It stared at her for several minutes before
sidling back into the forest. This was not the only sighting of the
famed Sasquatch in the area around Dulce.

Because of Valdez's unparalleled access, we were able routinely to "drop in" and interview some top-level members of the tribe. It was highly unusual to have such access, especially for a white person. One of the top three people in the tribe in the late 1990s was Wayne Gonzales (not his real name). Wayne is a stocky, well-built man is his early sixties with an impish and irreverent sense of humor. He is also a successful rancher and owns a sizable spread outside Dulce. Wayne told us of an encounter that he had with a Sasquatch on his property in the summer of 1993. In was during late afternoon. Wayne was sitting on his horse with three of his dogs nearby when he saw a Sasquatch break cover, running from the trees. The creature, which was covered with long brown hair, appeared to be running from an invisible pursuer. It was running on two legs like a human and it was running fast, passing just 150 feet away from Wayne. The creature was as tall as Wayne as he sat on his horse, or at least seven feet high.

The three dogs reacted to the creature's appearance by running under the horse, which was unique behavior for the dogs, to say the least. The creature then saw Wayne, since it had been looking over its shoulder on the opposite side, but it did not react to him. The creature ran swiftly toward a nearby hill and suddenly disappeared "into thin air." Wayne was certain that the creature did not run out of sight but actually ran through some kind of invisible "opening" before vanishing. The dogs stayed very close to the horse as Wayne rode to the top of the hill. Suddenly, the horse jumped several feet as if clearing an obstacle, but Wayne could see nothing. He noticed the dogs also jumping, but he could still see nothing.

Wayne's description is striking for two reasons when compared with the incidents on the Utah ranch: first, because the creature itself was seen in both New Mexico and in Utah and second, because Wayne's narrative encompassed an apparent vanishing of the creature into thin air.

A game and fish warden who had retired by the time we interviewed him around Dulce told us of his sighting of a strange, heavily muscled creature that walked across the track a mere fifty feet in front of him in the 1980s. The creature looked like a hyena but had a prominent, possibly even a bushy, tail. This creature was reminiscent of or similar to the brown, heavily muscled creature that attacked the Gorman horses in the corral in 1999 (see pages 190–92). One has to ask why two apparently nondescript areas are so prone to such a bewildering variety of creatures, unidentified flying objects, and otherworldly phenomena.

Sightings of bizarre orange balls were legendary in the area around Dulce. We spoke to dozens of witnesses who had seen these mysterious spheres. Many witnesses reported that they moved erratically in an abrupt, jerky motion. Two police officers once watched an orange orb hovering about twenty feet over a herd of cattle just a couple of nights after a cow had been mutilated a mile away. The two officers, who worked for the tribe, radioed Officer Gabe Valdez, who was sitting just across the valley in his patrol car. Valdez was watching the same orange orb from a distance. The object moved quickly as Valdez took off after it at 100 miles per hour on the narrow roads. Maybe it had intercepted the radio traffic and knew it was being hunted? Valdez harbored deep suspicions that these things were capable of sophisticated eavesdropping, but he had no idea who owned them.

As he approached the other two officers, the sphere had vanished, but Valdez saw a dark object move over his speeding patrol car. It was moving swiftly and climbing. Valdez was certain the orange object had simply extinguished its light so that it could avoid detection. It was heading north and was perfectly hidden except for the telltale round silhouette Valdez saw against the moonlit sky. Within seconds it was gone. Did the officers interrupt some animal mutilators? He would never know. But the orange object was big enough for several people. Were they the mysterious mutilators? Valdez had hunted the mutilators for decades and still had never caught them.

Only a fraction of these New Mexico sightings had ever seen the light of day. The mind-numbing assortment of objects—from huge, mile-wide metallic objects to several different colored orbs of varying diameters, to silver discs, and large one-hundred-foot-diameter domed discs—was reminiscent of the bewildering range of objects seen by the Gorman family and others in the Uinta Basin. Did the large number of objects have some significance? Was it indicative of different phenomena? Were the witnesses just hallucinating something weird in various ways? Was Dulce (and northeastern Utah) a testing ground for a whole slew of U.S. government secret projects?

CHAPTER 22
Other Hot Spots

No place in the world, other than perhaps Dulce, has experienced the sheer range of bizarre experiences that have been reported at the Skinwalker Ranch. Certainly few other reputed hot spots have been subjected to the same level of intense, long-term scientific observation as has the ranch. That said, there are a handful of other locations where similar phenomena have been reported over the years, places where the sightings of unexplained aerial objects are but a small part of a larger—and stranger—picture. If there is a lesson to be learned or a pattern to be discerned from what has occurred at the Gorman ranch, perhaps there are clues in the experiences of other families who have encountered the unknown.

On the eastern slopes of the Cascade Mountains in south-central Washington, residents have been reporting odd aerial objects for generations. Native American legends extend the stories back for hundreds of years. One particular region, composed of twenty-eight-hundred square miles of thick forests, farms, and the Yakima Indian reservation, is regarded as one of the most consistent UFO hot spots in North America. So many unexplained nocturnal lights and mystery aircraft were reported by residents in the late 1960s and early 1970s that a formal study was initiated at the request of J. Allen Hynek, the former Northwestern University astronomer who had served as chief civilian investigator for Project Bluebook, the U.S. Air Force study of UFOs that was formally terminated in 1969. The so-called Toppenish Study of UFOs in the area near the Yakima reservation was conducted by Hynek's associates in 1972, was updated in 1974 and 1975, and, most recently, was revised in 1995. The investigators documented dozens of UFO sightings and close encounters.

The people who live on or near the Yakima reservation didn't need a study to tell them that something strange had been going on in their sky. Investigators, then and now, say they are hard-pressed to find anyone in the region who *hasn't* seen a UFO. Personnel stationed in fire control towers say the sightings of orange balls of light became so routine that they would scarcely report them. Police officers, ranchers, government employees, and everyday citizens have seen not only orange balls of light that seem capable of performing impossible aerial maneuvers but also aircraft larger than jumbo jets that witnesses say could stop in midair, perform U-turns or zigzag patterns, then zoom away in the blink of an eye.

But in central Washington, as in northeastern Utah, the strangest events occurred on the ground, not in the sky. Writer Greg Long, whose book *Examining the Earthlight Theory* chronicles decades of UFO activity near the Yakima reservation, found a ranching family whose experiences mirror those of the Gormans to a remarkable degree.

Like the Gormans, the "Smith" family chose to live in a rural farming and ranching community. The fertile farmland near the Yakima reservation is known for its corn, sugar beets, apples, and timber, as well as for the quiet, country lifestyle that comes with the territory. Bill Smith, his wife Susan, along with their son and daughter, moved onto their property in 1966.

It wasn't until 1969 that the family experienced its first "paranormal" event on the farm. Susan Smith says she heard the family's dogs "throwing a fit." She responded and saw that the animals were barking ferociously at a boy who was walking along the road in front of the farm. The boy appeared to be Hispanic or Indian, was wearing blue jeans and a blue shirt, and made no sound. As he walked along, he passed behind a tree on the edge of the property, but he never re-emerged on the other side of the tree. After a few minutes, Susan ran to check on him, thinking he might have tripped or otherwise hurt himself. The boy wasn't there. He had simply vanished. Susan was so upset by the strangeness of the encounter that she ordered the tree to be cut down.

Family members began to hear disembodied voices of men, women, and children. At times, the voices spoke perfect English. Other times, the language was unknown to family members; they described it as "guttural." On one occasion, Bill and Susan returned home from dinner and heard the voices of little girls inside the house. The mystery girls sounded as if they were at play, laughing and giggling. As soon as Bill and Susan stepped into the home, the voices stopped. Bill recalls the night he heard a loud male voice that sounded as if it was coming from right outside his bedroom window. The voice proceeded to describe the physical characteristics and exact location of everyone inside the house. When Bill went to the window to peer out, the voice stopped. No person could be seen outside, and the family's usually alert watchdogs failed to notice any intruders.

In addition to the voices, the family grew accustomed to

hearing strange noises, including the sounds of footsteps inside the house, bumps and bangs on the walls and on the porch, a frequent electronic beeping noise that would come and go with no apparent source, and a hammering sound that persisted for more than a year. Bill described the sound as that of a metallic post driver hammering a post, over and over, into the ground. The sound usually began around dusk and would persist intermittently, sometimes until dawn. It seemed to emanate from above the ground and was heard by others who visited the farm. Whenever Bill or another family member would go out to investigate, the hammering would stop, only to start again fifteen minutes later.

Each family member—as well as visitors to the property—reported a number of poltergeist-type events. Knives propelled themselves out of frying pans, appliances jumped off hooks, and doors opened and closed of their own volition. Heavy objects mysteriously repositioned themselves. Tools disappeared, then reappeared in the same spot minutes later. A family friend who volunteered to help dig a ditch said he was repeatedly approached, in broad daylight, by what appeared to be the shadow of a man. Shadowy legs would walk up to where he was working and just stand there. The friend, understandably, was spooked and left the property.

During the years when all of these other events were unfolding, the family repeatedly experienced a range of UFO-type encounters. Almost from the beginning, the family saw weird lights nearby. Sometimes, the balls of light were yellow, ringed by an orange rim. Other times, the flying balls were red-orange. In most cases, they appeared to be intelligently controlled. On a few occasions, the property was flooded with an intense light that seemed to emanate from the sky, although no source could be seen.

In the summer of 1972, the family had perhaps its most spectacular UFO sighting. A brilliant light, described as brighter than the sun, seemed suspended above the Smiths' cornfields. The light would dim for a few seconds, then move a few hundred feet, then become more intense again. At one point when it dimmed, Bill was able to see the source—a large Zeppelin-like craft that he estimated to be the length of three boxcars. The airship was the color of oxidized lead, like an old battleship, and the family could see what appeared to be a row of portals in the side. The craft had no wings, fins, rudders, or obvious form of propulsion as it silently traveled over the cornfields at a speed of thirty-five miles per hour.

Bill had a fleeting thought to grab both his rifle and a Polaroid camera but says he received a telepathic message that the object had the family "in its sights" and that it would be foolish to take any hostile action. Bill froze and touched neither the gun nor the camera. It was his impression that the message had come from some type of advanced intelligence aboard the airship.

One day they found evidence that something landed on one of the lawns. The grass was crushed in a rectangle five feet long and two feet wide. Something else alighted on the family's automobiles on two occasions, leaving behind a weird pattern of circular markings, as if someone on crutches had walked on the hood and sides of the vehicles.

Because of the family's request for anonymity, it isn't known whether the Smiths still live on the property or whether these occurrences are ongoing. Investigators who have pursued the case have few solid ideas about what was behind the strange series of events.

The similarities to the Skinwalker Ranch in Utah are numerous—a rural family of four experienced both UFO- and poltergeist-type events over a long period of time, for no apparent reason or motive, and in the vicinity of a Native American tribal community. The events seem to have been orchestrated by some intelligence. This intelligence seemed to react to the emotional states of the family members. While the farm property appeared to be the epicenter of the strange events, there were many other manifestations that were witnessed and reported by many other residents of the region.

A remote ranch in rural Colorado was the site of events that even more strikingly resemble the activities at the Skinwalker Ranch. Once again, investigators have withheld exact details about the location of this ranch and the identities of the witnesses, but enough is known about this case to draw obvious comparisons.

The case was originally investigated by a team associated with the now-defunct Aerial Phenomena Research Organization, one of the leading UFO investigative groups of the 1960s and 1970s. Investigators included Leo Sprinkle, a psychologist from the University of Wyoming, anthropologist Peter Van Arsdale, and seismologist John Derr. Popular British UFO researcher Timothy Good publicized the results of their investigation in his 1991 book *Alien Contact.*

According to Good, the principal witnesses were a husband and wife named "John" and "Barbara," their teenage sons, and family friend "Jim," a former U.S. Air Force security officer. John, Barbara, and Jim pooled their resources to purchase a somewhat run-down ranch that had been vacant for some years. The property contains grazing pastures, wooded areas, and natural springs, and is said to be located in the general vicinity of ongoing military operations, although the name of the base has never been made public.

Over a period of four years, family members experienced a wide range of bizarre and frightening occurrences. It began with strange noises and electronic humming sounds that had no apparent source. The family came to associate the humming noise with UFOs. UFO sightings became almost commonplace on the ranch. In one dramatic instance, the witnesses say, a fleet of nine flying saucers landed in front of the ranch house. They also saw other craft of varying shapes and sizes floating silently past the property, both during the day and at night. Family members say they frequently heard the sounds of heavy footsteps in and around the ranch house, as if the place was haunted.

In October 1975, the family's teenage sons came across the remains of a cow that had been mutilated. They also spotted huge footprints that they later learned were made by a Bigfoot-type creature that was spotted on or near the property several times. The footprints suggested that the creature had followed the teenagers back from the woods after their discovery of the mutilated cow.

After a second animal mutilation on the ranch, Jim went to town to discuss the situation with lawmen. The deputy he contacted told him that there had been hundreds of mutilations in the area and that there was no reason to investigate them all since they were likely carried out by "extraterrestrials." Colorado, it should be noted, is where the first publicized animal mutilations occurred. Hundreds, perhaps thousands, of cases have been documented in the state, and they continue to this day.

Jim was unwilling to accept the explanation offered by the lawman. He and the others began to suspect some sort of military involvement in light of the ranch's proximity to assorted military activities. For a time, they also theorized that an unscrupulous real estate agent might have been sabotaging their ranch operation as a way to force them to sell the property. In time they would discard both of these suspicions.

The Bigfoot sightings continued. In one instance, a large, hairy humanoid pushed its way through a barbed-wire fence and chased after family members, who ran into the house. They returned later and found not only eighteen-inch footprints but also a tuft of hair on the wire, hair that was tested by a Denver biologist who could not identify the species. Another time, Jim saw one of the hairy humanoids running beside the corral and he shot it with his rifle. He says he definitely hit his target, but the creature barely flinched. He found no blood or other trace at the spot of the shooting. When he chased after it, he heard a weird sound that resembled a soft whine combined with an electronic beeping noise. Jim reported the shooting incident to law enforcement, which severely chastised him for using his gun. (In addition to family members, a total of twenty witnesses claim to have seen the Bigfoot creatures at the ranch, including friends, visitors, and employees.)

There were other sorts of intrusions as well. At times, family members found themselves paralyzed, unable to move. Barbara was knocked unconscious by a powerful and unseen force as she peered out the window at UFOs floating in front of the house. After coming home one night to find the children huddled in fear, Jim ran out into the yard and yelled a threat to blow up the property, arguing to the unseen intelligence that if his family couldn't have the property, then no one would have it. He says a response came in the form of a clear, loud voice, in what he described as stereophonic sound, that told him, "Dr. Jim, we accept."

The telepathic truce didn't last, however. Family members witnessed humanoid figures clothed in tight-fitting suits on the property. On one night, Jim awoke to find a nearly seven-foot-tall being in a space suit, complete with cosmonaut helmet, standing beside his bed. The being vanished before his eyes. Jim went to his lawman friend and talked about moving the teenagers off the property because they might be in physical danger. The lawman reportedly counseled him that while the phenomenon had sliced up cattle and other animals, and had certainly frightened residents of the area, no physical harm had ever come to any humans.

Another incident occurred one midnight as the family entertained a few out-of-town guests. All lights in the house went out. The group was gathered around a stereo and had been listening to the phonograph. When the power went off, a voice emanated from every speaker in the place, the stereo as well as the television. The voice told the group, "We have allowed you to remain. We have interfered with your lives very little. Do not cause us to take action, which you will regret. Your friends will be instructed to remain silent about us."

One of the guests, an electronics expert, dismantled the stereo, determined to find out how this could have happened. The radio receiver had been off when the voice was heard; only the phonograph was being used. The electronics expert was baffled. He speculated that some sort of transmitter might have been used, but such a transmitter would have had to be incredibly powerful and sophisticated.

The most dramatic episode occurred in January 1977. Jim felt a compulsion to travel to the top of a hill on the property. John's oldest son accompanied him. Months earlier, the hilltop had been burned by something that left a charred thirty-five-foot circle where nothing would grow. Once the duo arrived on the hilltop, they noticed a yellowish light in the trees. When they approached the light, they saw that its source was a metallic box. The box emitted a humming noise. Jim advised the teenager to stand back as he walked toward the box. When he got within a few feet, the humming sound changed to a louder, angrier tone, something akin to a swarm of enraged bees. Jim cautiously decided to walk the teenager back to the car, which was a short distance away. When he returned seconds later, the box was gone. (Jim's lawman friend later recounted his own encounter with a similar box that he spotted under a tree. Rather than approach it alone, he returned to the office and brought a fellow deputy back to the spot. Not only had the box disappeared, but so had the tree.)

Jim's hilltop mystery didn't end with the disappearing box.

Later that same night, he saw another light in the trees. He sent the boy back to the house for safety, then went to investigate. When he arrived at the source of the light, he saw two men whom he described as short, almost effeminate, with large eyes and blond hair, and wearing what appeared to be tight-fitting flight suits. Sixty feet down the hill, a dimly lit flying saucer rested on the ground. And in the shadows stood a Bigfoot creature.

"How nice of you to come," the strangers reportedly exclaimed in perfect English.

Jim says his encounter lasted about five minutes. According to Timothy Good, the strangers apologized to Jim for the inconveniences they had caused and assured him that "a more equitable arrangement" would be worked out. They told him several other things that he found to be insignificant but insisted that he should not repeat any of it. Jim had many questions but didn't ask any. He says he advised the strangers that the mutilations of animals might be unwise since it might draw too much attention to their presence, but they declined to admit any involvement with the mutilations. The strangers reportedly demonstrated their control of the Bigfoot creature by ordering it to pick up the blinking box. When it touched the box, the huge beast dropped to the ground. The strangers then emphasized that contact with the box could be lethal.

The two strangers stated that Jim's memory would not be tampered with, implying that they had the power to do so if they wished. They said that they would return to talk again. Jim felt it was time to go and walked back toward the house, his head spinning. He wondered if they were going to give him the cure for cancer, or a billion dollars, and he pondered what might be meant by "a more equitable arrangement." He was not convinced that they were space aliens and remembers thinking that somehow the government was behind the whole thing.

The possibility of military involvement isn't without some

foundation. During the period of the cattle mutilations, the family often saw military-type helicopters flying over the property. Jim believes that the helicopters he saw were too small to have carried out the mutilations because, he reasoned, whoever was responsible most likely hoisted the heavy animals into the air to perform the surgeries, then deposited the carcasses on the ground. In his estimation, small helicopters couldn't lift a twenty-seven-hundred-pound bull.

His curiosity was further aroused when he called the nearby military base to complain about helicopters that sometimes landed on his property. During his conversation with a colonel, Jim says he was asked what he thought about the cattle mutilations. The question seemed to come from out of the blue. Jim offered the opinion that perhaps UFOs were responsible. The colonel then admitted, in an amazing burst of candor, that the base was having its own problems with UFO intrusions. The colonel then asked if Jim had filed any complaints about his Bigfoot sightings! The officer confided that base personnel had received strict security instructions about how they should deal with both UFO incidents and Bigfoot sightings.

Such a conversation, if it happened, would seem to violate any number of military protocols. At least one critic of this case has argued that this odd, off-the-cuff chat strongly suggests that the officer was playing along with some sort of psychological exercise, mind game, or disinformation effort.

The team of APRO investigators told Tim Good that they believed the witnesses were being truthful in their description of these events. Psychologist Leo Sprinkle said he was initially skeptical of the case but that he viewed the principal witnesses to be "reliable and sincere." The other investigators agreed with this assessment. If it was all a mass hallucination, its long duration and daunting complexity would certainly put the entire affair in unique company.

While Tim Good and the original investigators have declined to make public any specific information about the location of the ranch, an article on the Internet sheds some light on the case. The paper, "UFOs: The Military Unmasked," was written in French by researcher Emannuel Dehlinger in 2003. It identifies the location of the ranch as Elbert County, a sparsely populated area southeast of Denver. Descriptions offered on the official Elbert County website seem consistent with what is known about the ranch. The two-thousand-square-mile county, home to a mere 22,254 people in 2003, is known for its ranches, farms, abundant wildlife, and sleepy country charm. What Elbert County doesn't have, though, is a military base.

That said, Elbert County abuts El Paso County, which is home to three large military installations, including Peterson Air Force Base, the hub of the USAF space command operations, home to the Twenty-first Space Wing, which is the sole organization within the USAF responsible for worldwide missile warnings and space control operations. The U.S. Army's Air, Space and Missile Defense Program is also headquartered there. What's more, Peterson AFB's elite "Team 21" works directly with NORAD's Cheyenne Mountain complex in protecting the United States from any intrusion or potential threat from the sky. When the residents of the ranch said that they had a vantage point from which they could observe military operations, could they have been talking about activities on test ranges controlled by Peterson AFB?

According to Jim, the military base he contacted was well aware of UFO incidents, animal mutilations, even Bigfoot sightings on the ranch. The colonel's astonishing and somewhat cavalier acknowledgment of these bizarre episodes led writer Emannuel Dehlinger to suggest that the military had orchestrated the entire saga. Dehlinger's paper argues that a covert military agency conducted a "psychological manipulation" of the witnesses by using electromagnetic waves somehow to induce hallucinations as well as to manipulate electrical devices. Dehlinger thinks a combination of advanced hypnosis and costumed actors may have convinced the ranch family that their encounters with the assorted beings were real. And Dehlinger argues that an artificial luminous projection system, perhaps based on plasma technology, could have been used to create the many UFO sightings.

Why would anyone go to so much trouble to scare the hell out of one unfortunate family? Dehlinger thinks the military wanted to take control of the property because it overlooked a strategic range that was being used by the air force. Sure enough, in 1979, the family gave up, sold the ranch, and moved away. Similar suggestions have been made about the Utah ranch, that a covert military group created an elaborate melodrama in order to seize control of the property, for whatever reason. And, as we know, the Gormans did move.

But the most glaring weakness with such an argument is that the military has no need for an incredibly sophisticated, long-lasting, and undoubtedly expensive charade. If the Pentagon determines that a particular piece of real estate is needed for strategic, security, or intelligence reasons, it can simply seize the land and deal with the consequences later. Such is the case with Nevada's infamous military base known as Area 51. In the mid-1980s, the U.S. Air Force commandeered eighty thousand acres of public land adjacent to the base in order to create a larger buffer zone. Armed security forces were positioned around the previously accessible acreage. Two years after the seizure, the Pentagon petitioned Congress for the official permission to do what it had already done. (Years earlier, the managers of Area 51 forcibly ejected a family from the privately owned mining operation because the mine site had a view of the secret base. The family had owned the mine for nearly one hundred years.)

Both the Utah ranch and its Colorado counterpart had been vacant for years before the new families moved in. If the military truly wanted the properties, it could have quietly acquired them during those years. It didn't need to produce a special effects extravaganza worthy of Steven Spielberg or George Lucas just to get some ranchers off their land.

CHAPTER 23
Vandals

By the end of July 1997, no scientifically useful data had yet been obtained at the ranch. As a part of a switch to a more proactive and less reactive strategy, the NIDS team decided to install a number of surveillance cameras near one of the hottest spots on the ranch. A series of six surveillance cameras were deployed in an area several hundred feet from the command and control center. The area was chosen because a series of dramatic events had occurred nearby, including the brutal mutilation and killing of a calf a few months before, the disappearance of six cattle, and the observations of the blue orbs and the famous orange holes over the cottonwoods. If any orange hole ever opened up in the same area of the sky, it would be captured in all its glory by the surveillance cameras.

Seven days a week, twenty-four hours a day, the cameras silently recorded everything in the sky and on the ground of a good portion of the ranch. The cameras were standard surveillance equipment, some of which had the ability to see into the infrared. We kept hard copies of the recordings and monitored them for unusual activity.

A year passed. The cameras picked up nothing unusual.

Then on July 20, 1998, Tom Gorman noticed that three of the cameras had stopped recording. He walked the several hundred feet out to the telephone pole to see if a lightning strike (of which there were several that summer) had disabled the cameras. What Tom found necessitated a fast telephone call to Las Vegas. I took the call. Tom told me that somebody had badly vandalized three of the surveillance cameras—the wiring had been forcefully ripped out. But the three other cameras were still running. The call provoked a fast trip to Utah for the research team.

The ranch was green and the vegetation was especially verdant that summer. The cattle grazed as usual in the large field nearest to the homestead. As I examined the cameras on the telephone pole, it was obvious that somebody was intent on inactivating the cameras. All three cameras had been mounted about fifteen feet off the ground, each camera facing a different direction, so that the full 360 degrees could be covered. From the three cameras, the video feed and the power feed came together and went underground at the base of the telephone pole. All the wiring had been neatly anchored to the pole by means of PVC tubing and also by heavy-duty duct tape. Each set of wiring from individual cameras had been separately wrapped in duct tape.

The PVC tubing now lay bent and twisted at the foot of the pole. All of the duct tape had been meticulously unwound from both the individual wiring and from around the telephone pole itself. Finally, the wiring itself had been dragged forcefully out of all three cameras. Anybody who has experience in wrapping wiring with duct tape and leaving it to bake in the summer sun will appreciate the enormous difficulty and patience required to unwind the sticky tape several times from individual wires as well as from around the pole. The marks of the duct tape on the damaged wiring as well as on the telephone pole were plainly visible. The tape was nowhere around. We searched the ground minutely for several days for several hundreds of yards radius around the pole for the missing duct tape but found no clues.

The next part of the investigation was to see if all three cameras went off line simultaneously, and for this we had to rerun the videotapes. We also hoped to obtain a clue regarding the perpetrators because it was possible the cameras might have recorded the unknown vandals during their approach to the telephone pole. The videotapes revealed that all three cameras had lost power almost simultaneously, at about 8:30 the previous night. This was just before it had got dark, so there would have been enough light to maybe catch the perpetrators in the act of vandalism.

Suddenly it dawned on us that one of the remaining three cameras, on a separate telephone pole two hundred feet away, was still working. Better still, it had been pointing directly at the vandalized pole during the incident. There was a mad scramble to retrieve the videotape and we waited in breathless anticipation as we played back the video. Sure enough, the camera had faithfully recorded everything; there was still plenty of light out as the time stamp rolled toward the fateful 8:30 time when presumably the recording would reveal who had damaged the other cameras.

We watched dumbfounded as the time stamp continued past 8:30 and revealed no unusual activity. Carefully, we replayed the videotape, checking to make sure we had the correct time. The videotape time stamp was precise. But we could see nothing on the footage other than the pole itself and the cattle peacefully munching grass in the pasture beyond. Each time we replayed the tape, the more certain we became that we had the correct time. This made no sense. Whoever had yanked the wiring out of the cameras should have been in plain view on the tape. Nothing whatever disturbed the pristine stillness. Unfortunately, the resolution was not good enough for us see the thin wiring in any detail, but it should have afforded us a glimpse of whoever had ripped the wiring to pieces.

Completely flummoxed, we took all of the videotapes back to Las Vegas in the hopes that digital enhancement might give us a clue as to what had happened. After multiple rounds of digital enhancement, the puzzling incident came into even sharper focus. The resolution became good enough to see the tiny red lights on the bottom of each camera suddenly lose power at precisely 8:30 P.M., confirming that the cameras had actually lost power while being taped. Yet the enhancement provided no clue whatsoever as to who or what had so forcefully damaged the equipment.

The combination of fastidious attention to detail in removing every shred of duct tape from the wiring, while at the same time ripping out the wiring from the camera junction boxes was disturbing. It reminded me of the same eerie combination of brute force and finesse that had been displayed in dismembering the calf a little over a year previously. And interestingly, both of these incidents had happened less than fifty yards from each other in the same field, in daylight, and in full view of witnesses in one case and cameras in the other.

The three remaining cameras on the ranch continued their surveillance. Another eight months passed.

Tom Gorman was still working as ranch manager in April 1999, hoping to get some "closure" on the bizarre incidents that drove him off the property. It had been three years since the family had vacated the property that they had once loved. He waited silently for Ellen to jump back in the pickup after she had padlocked the gate behind her. He drove slowly down the familiar track leading to what used to be his homestead. It was a bright, sunny afternoon.

"What is that?" Ellen muttered, looking puzzled. Tom squinted west. The sun was in his eyes, but he could make out the clouds of dust coming from the corral about a quarter of a mile away. Tom kept a couple of his horses in the corral just in case he had to saddle up to retrieve any errant cattle that had broken through the fence line—an all too common occurrence in the past three years. The cattle had a nasty habit of stampeding en masse through the fence line. This usually meant hours of work for Tom to ride the cattle down and to herd them back onto the proper land. He had understandably grown tired of driving the twenty-five miles back to his place, putting the horses into the trailer, and then driving back to the ranch in order to saddle up and play cowboy. It was a lot easier simply to have a couple of horses in a corral at the ranch on standby.

Tom was puzzled as he drove. He could see the horses moving quickly around the corral in the distance. Their wild kicking generated clouds of dust. The Gormans were about three hundred yards away when it became obvious that the horses were not alone in the corral. Amid the clouds of dust and spooked horses, Tom could make out a reddish brown blur that seemed to be running around the narrow space. He mentally made a note to himself to give a good thrashing to whatever dog was doing this to his horses. Panicking the horses like this might be good fun for the dogs, but it left the horses stressed for days.

As they drove closer, the scene became clearer. "That is no dog," Ellen said as she leaned out the open window. They were two hundred yards away and she had a much better view of the corral out the passenger side of the pickup. She was right. Tom slowed to get a better look. Although the clouds of dust were still being kicked up, he could now see the perpetrator.

It was a big animal, very heavily muscled, with short legs. It seemed to have the shape of a hyena's body, but it had a bushy tail! "What the hell is that?" he muttered. The creature was plainly hunting his horses but did not seem intent on causing serious harm. It would lunge at one of the horse's legs, looking to bite it. The horses would then kick out and gallop around the corral.

The animal's big red bushy tail reminded Tom of an exaggerated fox tail. But the rest of the body was all wrong for a fox. The animal looked and moved like a hyena, but its head more resembled a dog's. It had short, stubby legs like a boar, as it made sudden rushing movements across the corral. They were forty yards away now and dismounting from the vehicle.

Tom and Ellen now had a very good look. This was like nothing they had ever seen before. It was definitely not a dog. And it definitely was not a fox. Nor a hyena. Its reddish coloring was all wrong. He figured the animal weighed about two hundred pounds. However, his immediate concern was for his horses. Tom slammed the pickup door and took off running.

The creature suddenly stopped the hunt and turned to look at Tom as he sprinted toward the corral. Instantly, the creature took off in the opposite direction, jumped nimbly through the metal bars of the corral and headed up a slight incline. Tom reached the side of the corral, no more than fifty yards from the fleeing animal. He was determined to see where the animal was heading. It ran quickly up the slope away from the corral with Tom in hot pursuit. Then it was gone.

Tom stopped. The creature had vanished into thin air. It was open ground. The creature was far too big to have vanished down a rabbit burrow, but Tom would have seen the large bushy tail anyway. Within a minute he had reached the place where the animal had disappeared before his eyes. There was nothing there.

The ground was too hard for tracks. Tom caught the distinctive smell of wet fur in the air. It reminded him of the rank smell of a wet dog.

After searching fruitlessly for a few more minutes, Tom hurried back to his horses. Both of them had bloody hocks. They were severely scratched but not seriously hurt. He figured the animal could have inflicted much more serious injuries on the horses if it had meant business.

After 1999, there were other sightings of the large, reddish hyenalike creature, once by a ranch employee and also by a local man a few miles from the ranch. In both cases the creature fit Tom's description of the original two-hundred-pound animal.

Tom had seen other strange animals on the ranch, of course, like the large wolves, one of which was bulletproof, that had haunted the ranch for several weeks after the family had moved onto the property and then had disappeared into the mists, never to be seen again. Tom also told me about some tiny bright red birds that had suddenly appeared. For a few days the birds, about the size of a wren, fluttered around a couple of trees near one of the abandoned homesteads and then vanished, forever. The birds' fiery red color together with their tiny size made them seem more like tropical birds, not indigenous to northeast Utah. Tom also told me about huge spiders he had seen in the same area around the abandoned homestead. Just like the birds, he saw the large spiders for a few weeks, then never saw them again.

Other locals saw strange animals as well. In October 1998, a man and his wife were returning to their home. As they drove along the narrow roads about three miles from the ranch, the wife spied some movement in the field next to the roadway. It was dusk, but there was still sufficient light to make out a humanlike figure running across the field. As the couple watched, the figure kept up the pace of an Olympic sprinter over hundreds of yards. They could not make out its features, but it looked like a dark, very muscled man running smoothly and effortlessly at an unbelievable speed. The creature or man was running in the direction of the ranch. What amazed the two witnesses was the smooth and rapid pace the running figure sustained. They watched it as it disappeared gradually from their field of view.

Throughout the years, the surveillance cameras had continued recording data, and with the exception of maybe a dozen instances of fast-moving meteorlike objects and suspicious aircraft activity (possibly due to drug smuggling), we managed to obtain no sustained evidence of anomalous phenomena. We still mounted regular field trips to investigate activity on the ranch but noted nothing of any consequence during our visits. In addition, beginning in 1999, two people began living on the old Gorman homestead, with instructions to keep eyes wide open and to report anything out of the ordinary to headquarters in Las Vegas. But the years rolled on with progressively fewer and fewer incidents. Altogether we logged hundreds of night watches, but no sustained activity occurred with anything like the dramatic flair of the summer of 1997. By 2004, when I left NIDS, it had been several years since anything of note had happened on the property.

Tom Gorman had warned NIDS when we moved the giant observation trailer onto the property that our approach was much too heavy-handed. The flurry of activity and the large disruption that marked our arrival seemed to be a turning point in the intensity of the activity. The NIDS team never approached the kind of stealth that Gorman had advised. Instead, Gorman frequently implied that we had acted like bulls in a china shop. (We were, after all, hunting a quarry.)

Between August 1996, when NIDS acquired the Gorman property, and March 1997, not a great deal of activity occurred. Then with the dismemberment of the calf, all hell had broken loose. In May 1997, a series of unexplained phenomena followed the installation of a large wire enclosure for the sentry dogs. Between June and August 1997, the phenomenon was present every time scientific personnel were. This included the sightings of balls of light in June followed by the mysterious, apparently telepathic transmission to the physicist on the scene. During July, there were multiple sightings of balls of light and other light phenomena, all in the same area of the ranch. In August, the red ball incident occurred, and then a few days later that creature crawled through the hole. That incident marked the end of the period of intense activity in NIDS' presence.

Afterward, the phenomenon became much more fleeting. Did it lose interest? Or was there something now missing from the engagement? Perhaps it was the level of emotion that the Gorman family had provided in spades but was missing from the scientific team. The stress level in the family was unbelievably high. It was palpable. The Gormans did not interact with the phenomenon because they wanted to; they simply had no choice. In contrast, the NIDS scientific personnel were there by choice. They carried with them an attitude of cool detachment. There was almost an aggressiveness in the pursuit of the phenomenon that may have psychologically turned the tables, assuming of course that a consciousness was involved.

At the risk of inviting strong disagreement from other NIDS scientists, one could ask the question: Could NIDS have approached this research differently? Did our preoccupation with measurement reach a point of diminishing returns? It could be argued that our research was overly focused on one question: Is this real, or is it an artifact of (a) researchers' fugue states, (b) malfunctioning equipment, or (c) natural causes? In short, did NIDS leave any stones unturned? Could we have more emotionally engaged the phenomenon without sacrificing scientific objectivity? Was the period from March to August 1997, during which the activity appeared to intensify dramatically, some kind of test designed to assess the scientific team? And did NIDS fail the test? Was some kind of emotional engagement expected or needed in order to deepen the dialogue? Was "contact" of some kind being offered? Did NIDS' strict adherence to scientific protocols get in the way?

We may never have answers to these admittedly speculative questions.

PART III

AFTERMATH
AND HYPOTHESES

CHAPTER 24
The Media

R umor, innuendo, and wild speculation are the enemies of any scientific investigation, especially one that is focused on such unusual phenomena. Since the arrival of NIDS at the Gorman property, considerable efforts have been made to limit public disclosures about unusual events observed on the ranch. The policy was adopted for very practical reasons, but the secrecy has contributed to the creation of a vibrant and imaginative mythology about the ranch, a mythology that has taken on a life of its own. An ambitious sociologist could probably write a PhD thesis about public reaction to the various rumors that have sprung to life. It is a textbook example of how scraps of information can be weaved into a rich tapestry of nonsense and hyperbole.

A lot of what happened on the ranch before the arrival of NIDS is no secret. Tom Gorman, exasperated by the loss of his animals and by the psychological toll on his family, decided to go public in 1996. He related some of the details to reporter Zack

Van Eyck of the *Deseret News*, Utah's second largest newspaper, but Tom didn't come close to revealing everything that had happened, perhaps sensing that the full picture would be tough for anyone to swallow. The Associated Press was quick to pick up Van Eyck's excellent story, which pointedly did not belittle Tom or ridicule his experiences.

A photo of Tom standing beside the strange scooplike depressions in his pasture was distributed to newspapers all over the country. Radio networks also carried the story. The extensive coverage prompted the arrival of a few enterprising UFO investigators at the ranch. Two of them, Ryan Layton and Chris O'Brien, received permission from Gorman to stake out the property on several nights. Other would-be investigators were rebuffed.

Being a private person, Tom was uncomfortable with all of the attention, but he took the plunge into media-infested waters because he wanted the bizarre activity to stop. The loss of so many cattle had edged him closer to financial ruin. If military or intelligence agencies were doing these things, he reasoned, perhaps the press scrutiny would force them to back off. If not, maybe someone out there could help him find some answers.

The publicity caught the attention of NIDS, which quickly bought the ranch and set up a monitoring program. Not surprisingly, the arrival of NIDS scientists in a small, rural community generated its own media attention and spurred considerable, if subterranean, concern among residents. Who were these outsiders, and what were they really up to? Although NIDS didn't want any additional attention focused on the ranch, the decision was made to use the media to quell rampant rumors and to appeal for cooperation. NIDS investigators knew that they needed the trust of area residents so that people would talk to them about anomalies they had witnessed in the past and about any new developments that might be relevant to their study. This was especially true regarding animal mutilation cases. If any cows were carved up in the region, NIDS needed to know as quickly as possible so that physical samples could be collected and analyzed before decomposition obliterated any useful clues.

Key members of the NIDS organization granted interviews to a few Utah newspaper reporters. Zack Van Eyck wrote two lengthy articles in which NIDS personnel assured the public that their activities at the ranch were strictly scientific and aboveboard, that there was no truth to the rumors that they had made contact with either extraterrestrials or "lizard people," that NIDS was prepared immediately to investigate any animal mutilation cases at no cost to the ranchers, and that scientific protocols required a certain amount of confidentiality. The interviews tried to discourage further intrusions by avid but uninvited outsiders.

"We know so little in terms of what the overall scope of these phenomena are that it's just embarrassing to try and make some conclusions at this point," a NIDS spokesman told reporters. "Imagine that you have a phenomena that is very selective as to how it exposes itself and to whom. So if you have a tailgate, football stadium–type of atmosphere and everybody's got hotdogs and hamburgers and they're barbequing and waiting for the UFOs to come down, I don't picture a continuation of the activity."

This seemingly innocuous statement was pregnant with clues about the nature of the phenomena being studied, but the hints sailed over the heads of most who read them. After the articles appeared in print, NIDS initiated a news blackout. No further interviews were granted about activities at the ranch. No outsiders were allowed to enter the property.

For the next six years, a cone of silence surrounded the ranch. Dozens of unusual events were observed. Animals were mutilated. Mysterious aircraft appeared and disappeared. Gunshots were fired at unknown creatures. And "the entity" manifested itself in ways that challenged and bewildered the NIDS team, but none of these events were made public. There were occasional attempts by journalists and UFO enthusiasts to find out what was going on. Curious UFO research groups organized campouts and sky watches on the edges of the property, hoping to catch a glimpse of the rumored activity and to gauge what methods were being used by NIDS.

Public interest in the ranch was piqued in 1997 after a national science magazine published a specious and inaccurate article alleging that Nevada's ultrasecret Area 51 military base was moving to Utah. The article prompted speculation that the imaginary relocation of Area 51 to rural Utah was somehow related to the presence of NIDS. This assumption was flatly ridiculous. Rumormongering aside, the plain fact is that the outside world had no idea what was going on within the confines of the property.

This changed abruptly in 2002. George Knapp had maintained a working relationship with NIDS since its inception and had earned the trust of principal figures in the organization. The CEO of NIDS had shared, on a confidential basis, incident reports and a comprehensive chronology of ranch incidents and, after some prodding, allowed Knapp to visit the ranch and to bring along television cameramen Eric Sorenson and Matt Adams on different weekends in the spring and the fall. The excursions represented the first visits by journalists since the ranch had been purchased by NIDS years before.

In late 2002, Knapp received permission from NIDS to write an account of the ranch activities for publication. The organization reasoned that the publicity might generate new leads about other places around the world that might be experiencing similar levels of paranormal activity. It accomplished that and more. A two-part article, "Path of the Skinwalker," was published in subsequent issues of the *Las Vegas Mercury*, a weekly alternative paper owned by the *Las Vegas Review Journal*, Nevada's largest newspaper, but the text quickly spread far beyond Las Vegas newspaper racks.

"The articles generated a great deal of buzz, locally and internationally," said the paper's editor, Geoff Schumacher. "Soon after they went up on our website, people across the globe were hitting them. Most of these people had never heard of the *Mercury* before. Even weeks later people were still seeking them out."

The articles were reprinted in UFO magazines in the United States, England, and Brazil, were translated into French and Portuguese, and prompted other paranormal publications to crank out their own articles, most of which borrowed freely from the *Mercury* stories without attribution. Tabloid TV shows headed to Utah to produce segments about the ranch, even though they were not allowed on the property. NIDS was besieged with requests for interviews and further information.

On the Internet, a discussion group was formed to chat and share ideas about the ranch. Initially, the participants expressed amazement and great interest. But the tone of the discussion soon changed and members of the chat group began to focus on dark rumors. Some participants argued that the study of the ranch might be some sort of disinformation effort being promulgated by government intelligence agencies. NIDS was described as a "shadowy" organization that likely had ties to the CIA and might be funded by proceeds from drug smuggling or casino gambling.

One UFO researcher proclaimed on the Internet that NIDS board members were part of a secret cabal known as the Aviary. Members of the cabal supposedly had their own birdlike code names such as the Owl or the Penguin. The mission of the sinister Aviary, according to the imaginative writer, was to control all information about the presence of extraterrestrials on Earth. (The Aviary had been, and continues to be, a staple of conspiracy buffs.) Another allegation was that NIDS scientists had acquired ET technology on the ranch and were trying to exploit it for their own nefarious purposes. An underground UFO newsletter printed an opinion that NIDS researchers had been visited by a spaceman who had revealed the secrets of "free energy." Still another writer opined that the ranch might be a laboratory for secret genetic experiments involving mutilated cattle. And other armchair experts theorized that NIDS might be trying to create the underpinnings of a new global religion as a tool for manipulating the masses, presumably for unholy ends.

Those researchers who had visited the ranch prior to the arrival of NIDS were flabbergasted by the revelations in the *Mercury* articles. They found it hard to believe that Tom Gorman had not told them everything. Some publicly chastised NIDS for its reluctance to release more details. They argued that whatever happens on the privately owned ranch should be considered public information and should be disseminated immediately.

UFO researchers in Utah were also angered by the *Mercury* articles, one of which had appealed to "saucer enthusiasts and UFO nuts" to avoid further trespassing on the ranch. Members of UUFOH, the Utah UFO Hunters, an enthusiastic and well-meaning group, assumed that the description had been aimed at them. They vented their anger and dismay on the organization's website. While acknowledging NIDS' request that no outsiders do anything to interfere with the delicate give-and-take that was under way, they quickly organized new expeditions to the perimeter of the ranch. In retrospect, their curiosity-driven reaction is certainly understandable.

In the summer of 2004, one other ranch-related controversy surfaced, and it carried with it the darkest of implications. A brilliant but maverick physicist known to have extensive contacts in intelligence circles and the UFO community suggested on the Internet that one or two NIDS staffers had been found murdered on the ranch and that the slayings had been covered up. The implication was that an unknown force had gutted the unnamed researchers and that this might signal the beginning of an "all-out war with the extraterrestrials." The physicist considered this to be a matter of national security, and he invoked a comparison to the alien-invasion movie *Independence Day*. He further asserted that the only reason he had shared the sketchy information with others was so that he could find out whether it was true. In subsequent missives, he stated that the original source of his information was a member of the NIDS board. Needless to say, there is not a word of truth to this rumor. With the exception of nonspecific severe headaches, profuse nosebleeds, and minor contusions, no serious physical harm came to any NIDS staffer or to anyone else on the ranch, and it is ridiculous in the extreme to suggest that alleged crimes of this magnitude would or could be covered up.

For NIDS, the publicity and controversy from the skinwalker articles represented a double-edged sword. Overnight, the ranch became world famous. NIDS was deluged with letters, phone calls, emails, and interview requests. There were several intrusions and attempted intrusions at the ranch, even though the phenomena on the property went into apparent hibernation. But there was a payoff, of sorts. NIDS received a great deal of information about other, eerily similar hot spots, in North America and around the world.

The basic problem remained, however: how to explain what had been seen and experienced at the Utah ranch?

CHAPTER 25

Hypotheses

The eight-year program of intensive surveillance using equipment and PhD scientists makes the Utah ranch arguably one of the most closely studied anomaly hot spots in history. Putting the results of this investigation into a proper context requires a considerable broadening of the picture, however. The cruel truth of the matter is that the wide sweep of events and anomalies experienced by the Gormans and by NIDS investigators does not fit into any of the well-known subject areas that traditionally define anomalies research. Rather, the ranch experiences appear to overlap all of them.

Some of the events seemed to reek of poltergeist activity; some appeared to be paranormal in nature; some exhibited behavior reminiscent of UFO-like objects; and there was a smattering of cryptozoology-like events thrown into the mix. Were UFOs flying over the Utah ranch, or was the ranch haunted? And according to local lore, sometime in the 1970s after a series of unexplained happenings on the property, the ranch was even visited by a strange individual wearing a black suit and driving a brand new black Cadillac with tinted windows, an event that is reminiscent of the so-called Men-in-Black episodes. Brand-new Cadillacs were a rarity in the Uinta Basin in the 1970s and remain so today.

In short, it is very difficult to encompass all of the events reported at the ranch into a single tidy discipline. This is no semantic discussion because there is considerable mutual suspicion, and even hostility, among the groups who study UFOs and those who research mystery animals and those who study hauntings and poltergeist activity. Which particular research data from the Utah ranch does one have to discard in order to fit the rest into a tidy box? Or should we just ignore all of the tidy boxes? This kind of discussion becomes surreal, and even ludicrous, when one considers that most of mainstream science does not recognize the legitimacy of *any* of these research areas.

In the future, an adventurous sociologist might consider writing a paper that examines the "caste" system in anomalies research. The "nuts and bolts" UFO research people regard the "psychosocial" UFO researchers with disdain. UFO researchers in general regard the cryptozoologists with contempt. Cryptozoologists who embrace the possibility of a paranormal connection to Bigfoot sightings are generally viewed with derision because of the prevailing view that Sasquatch is an undiscovered primate species, not an interdimensional playmate of alien beings. Likewise, the paranormal researchers view the UFO researchers with disdain, while the ghost hunters keep their distance from everybody else. And all of this hostility and contempt is a vain and so far unsuccessful attempt to earn a small measure of respect and acceptance (and maybe funding) from mainstream science, a lofty but unlikely goal.

So how does one go about modeling the range of more than one hundred phenomena that occurred on the ranch between

1994 and 2004? The events the Gormans and the NIDS investigative team experienced were mostly isolated incidents that were transient, opaque to interpretation, and rarely or never repeated. Many of them represented very physical intrusions, including the killing and mutilation of a young calf, the "transfer" of four two-thousand pound bulls into a small trailer, scoop marks removing a few hundred pounds of soil, the destruction of three video cameras, injuries to the Gormans and to neighbors, and much more.

Other incidents involved events that were invisible to the human eye, such as the "creature" crawling through the hole in August 1997 and the "black object" seen by the physicist in June 1997, both of which were detected in the infrared region of the spectrum but not the visible. Tom Gorman witnessed an intense splashing as a result of some thing or some creature running at full speed along the canal on the ranch, but the actual creature that gave rise to the splashing was invisible to him. This happened in daylight as Tom sat on his horse only a few yards away. The "Predator" incident of early 1996 seemed to indicate a creature or an object that was barely visible but mostly camouflaged; while in April 1997 something invisible drove or ran through the herd of grazing cattle and parted them like the Red Sea. Whatever accomplished this conjuring trick appeared to profoundly affect a compass from a distance but was otherwise invisible.

An assortment of red orbs, blue orbs, and white orbs was observed. They appeared to be light forms, but they certainly had physical effects. The blue orbs acted like well-oiled rheostats by dimming the lights in the home as they flew past the windows. The same blue orbs were closely associated with the deaths of Gorman's three favorite dogs, presumably by incineration. Both Ellen and Tom were overcome by a fear that transcended everything in their previous experience as a blue orb hovered over them and seemingly stimulated the fight-or-flight reflex in their brains. The red orbs directly caused the death, injury, and abortion of several cattle in August 1997. The *chupas*, as they flew through the cattle, seemed to be associated with the mutilation and disappearance of both Gorman-owned and NIDS-owned livestock.

NIDS' attempts to measure this range of bizarre activity were unsuccessful, if success is defined by data published in a scientific journal, or even data that could be presented to a panel of mainstream scientists. The weird animals observed in late 1994 and in March 1997 felt—and smelled—like real physical animals, but their response to bullets fired from high-powered rifles was unquestionably different from that of any creatures known to science. I personally witnessed the shooting of a large catlike animal from a tree. I was forty to fifty yards away and am certain the bullet hit its target. I was just a few yards away when an unknown doglike animal weighing an estimated four hundred pounds was pierced by bullets fired by an experienced marksman, yet no body and no blood were found after both of these incidents. The only physical evidence was a single claw mark left in the snow. The saga of the bulletproof wolf is likewise difficult to interpret. In short, after several years of Gorman family trauma and of focused NIDS investigation, we managed to obtain very little physical evidence of anomalous phenomena, at least no physical evidence that could be considered as conclusive proof of anything. This was in spite of hundreds of days of human monitoring and several years of camera surveillance.

The plethora of tantalizingly short-lived phenomena gives rise to a simple question. The witnesses were exposed to a variety of apparently unconnected incidents, many of which were difficult to explain, but all of which raised more questions than answers. The central question is: Did all of this varied activity on the ranch originate from the same source? Or was the ranch a "Grand Central Station" where multiple phenomena intersected, all totally unrelated?

There are several hypotheses that have traditionally been used to explain anomalous phenomena, particularly of the unidentified flying object variety, of which the ranch certainly had its fair share. We will briefly examine each of these hypotheses in turn.

HOAX

Did the Gormans make it up? Could they have hoaxed the whole thing? It's true that the NIDS investigators took the Gormans at their word for those events that occurred prior to 1996. Skeptics might argue that the family made it all up, perhaps because they wanted to extricate themselves from a financial disaster.

We mention this possibility solely in the interests of thoroughness. First of all—and this is an admittedly subjective assessment—the Gormans are rock-solid, firmly grounded, honest people. There is no hint in their background of anything shady or questionable. The kids were excellent students, the parents were happily married, and the purchase of the ranch represented the culmination of a lifelong dream. The onslaught of bizarre activity did eventually take a toll on the family's reputation but, in light of what transpired, this isn't entirely unexpected.

Tom Gorman did not invent the long-standing reputation of the ranch. He didn't even know about it until it so abruptly intruded into the life of his family. He obviously could not have manufactured the stories told by the Utes. Besides, many people outside the Gorman family, including neighbors, law enforcement personnel, reporters, and assorted visitors, witnessed the incidents that occurred prior to the arrival of NIDS. And dozens of the most disturbing incidents were personally witnessed and documented by NIDS scientists and researchers. The Gormans did not—and could not—manufacture them.

Did the Gormans crave media attention? None of them had

the slightest interest in talking with the media or attracting attention in any way. By all accounts, they tried to cope with the inexplicable activity on their own terms and were embarrassed even to discuss it privately with others in the community. Only once did Tom Gorman talk to local reporters as a last resort, and then only reluctantly, hoping that someone might see the story and offer some kind of help. (Someone did—NIDS.)

In spite of everything that happened, Tom Gorman did not want to surrender his dream, and he departed the ranch with extreme reluctance. Although the family has moved to another state and has started a new life, we know that Tom still harbors a gnawing curiosity about what really happened during his years on that accursed patch of land. Today, years after they moved out of Utah, the Gormans don't grant any interviews and don't want their new neighbors to know about what happened.

In short, though it is impossible to rule out any individual incident as a hoax, especially those in which NIDS was not involved, it is highly unlikely that all of the strange events at the ranch were somehow manufactured by the family.

DELUSIONS

The scientific community usually rationalizes eyewitness reports of anomalous phenomena by assuming that the witnesses are operating from a delusional framework. Because of the popularity of this explanation among scientists, NIDS took it very seriously and spent a lot of time examining possible psychiatric explanations for the events that its investigators had witnessed. We reviewed the literature on stress-induced fugue states, on paraphrenia, and on the folie-à-deux phenomenon, as well as the vast literature on the interplay among hallucinations, psychopathology, and anomalous experiences.

We found the paraphrenia hypothesis particularly worthy of

examination. Note that we used the term *paraphrenia* as a catchall for a multitude of nonorganic psychoses, which is the basic European definition of the term, rather than the U.S. definition of late-onset schizophrenia. An individual can be paraphrenic in a narrow slice of life while maintaining high function in the remainder of his or her life. Thus, by definition, "paraphrenic perception" on the ranch would be very difficult to spot even by an objective observer. In fact, any differences between "normal" observation and slightly abnormal processing of information on the part of scientific observers might have been too subtle to catch. But suggestions that we conduct a battery of psychological tests—including Myers-Briggs Type Indicator, Minnesota Multiphasic Personality Inventory, indices of suggestibility, fantasy proneness, and more—on all NIDS field personnel was never implemented.

NIDS carefully investigated the wide variety of eyewitnesses to the incidents at the ranch and interviewed neighbors and locals who also experienced very similar events. Although in addition to psychological profiling, multiple discussions took place on the need to monitor stress hormones and other blood chemistry profiles of all researchers on the ranch, the initiative never went beyond the discussion phase. Such data might have contributed to supporting or eliminating the fugue hypothesis. Nevertheless, we found no common thread that might point to an underlying psychopathology among the various witnesses from different locations and backgrounds. It is unlikely that fugue states, paraphrenic delusion, or groupthink were involved in all of the occurrences, although isolated episodes are difficult to refute on these grounds, particularly those in which only a single individual visually perceived and reported an event while a nearby colleague could see nothing.

NATURE

Could a natural phenomenon be responsible for the events? Could the presence of either a seismic, an electromagnetic, or an environmental variable have induced consistent hallucinations in the human brain and altered the perceptions of both the Gormans and the NIDS investigators? NIDS spent a lot of time investigating and testing this hypothesis.

The best-known theory of environmental variables affecting human perception and giving rise to anomalous phenomena is the Tectonic Strain Theory (TST) proposed by Michael Persinger, John Derr, and others. These researchers have published widely on an alleged geographical and temporal relationship between mild tectonic or seismic stress and reports of anomalous aerial phenomena, including, but not limited to, earthquake lights. The TST holds that electromagnetic energy released from seismic stress sometimes manifests as light (the so-called earthquake lights) and also may interact with the human brain, perhaps as VLF (very low frequency) or ULF (ultra-low frequency) radiation to cause hallucinations, particularly in the temporal lobe. The former "direct effect" of earthquake lights has been seen and reported in conjunction with and in the vicinity of seismic activity, but the latter theory of an effect on the brain is little more than informed speculation.

Persinger and Derr have invoked the TST to explain the long history of UFO activity in the Uinta Basin. Therefore, the events at the Utah ranch provided an excellent opportunity to test the TST. NIDS attempted to correlate the timing and locations of seismic activity, using both United States Geological Survey and Utah State University seismic databases, in a three-hundred-kilometer radius of the ranch over a ten-year period with events reported by both the Gormans and NIDS investigators. But we found no such temporal or geographical correlations.

We also looked at a number of other possible environmental

variables, such as the possibility of ingested hallucinogens. But we found no correlation with local drinking water, since in most cases NIDS researchers consumed only bottled water. A close study of the flora on the property also failed to reveal an abundance of plants known to contain psychoactive substances. In the event that hallucinogens may have been windborne, we examined local wind direction data for a period of ten years and found no correlation with the timing of reports of anomalous events at the ranch.

Finally, we mapped magnetic field anomalies, using published aeromagnetic surveys as well as taking dozens of field measurements of magnetic field flux on the ranch ourselves—not to mention gravitational and other incongruities around the Uinta Basin—but all these efforts failed to correlate with the events observed on the ranch.

ADVANCED TERRESTRIAL CIVILIZATION

This hypothesis would have us believe that sometime in human history, maybe during the biblical era, during the Middle Ages, or even just prior to the Nazi era, an isolated group of humans gained access to advanced technology and have controlled world human affairs for their own ends ever since. To cover their activities, these humans are said to engage in deceptive simulations to persuade people that benign extraterrestrials are visiting Earth. Their presence on the Utah ranch would imply a covert program of unknown purpose, possibly a tiny but inscrutable part of a larger scheme fitting their overall global agenda. But if they are humans, they would still think mostly like us and make mistakes like us. Some puzzling lapses on the part of the phenomenon, including several instances when Tom Gorman was apparently able to hide on the property while a bright light searched for him, lend weight to such a hypothesis.

One anecdote in particular deserves elaboration. Gorman

was gathering hay one late summer evening in 1995. He noticed a bright light that appeared to be "watching" him while hovering above the ridge. He saw such objects often and was accustomed to being watched in this manner. For whatever reason, on this particular evening Gorman grew impatient, threw down his pitchfork, and ran in the direction of the bright object. His reaction seemed to catch the light by surprise. The object darted down out of sight behind the ridge. Gorman quickly seized the opportunity and dove behind a nearby hay bale and burrowed out of sight. Then he waited. Sure enough, within a few minutes the brightly lit object flew low over the field and began flying back and forth. Gorman sensed it was looking for him but could not find him. After several minutes of quartering the area, the object flew toward the ridge and Gorman climbed out from his hiding place and began shouting and taunting the object. Gorman said it blinked several times on and off and then flew away.

The incident left Gorman feeling elated because psychologically he knew he had won a small battle in this long war. He had succeeded in making a fool out of the probe or drone or whatever the bright light represented. Gorman thought he had also exposed the fallibility of their sensing system. The actions of the probe, as it flew around the field at low altitude, appeared to indicate it was not sophisticated enough to spot Gorman as he hid in the hay bale. Only much later did Gorman consider the possibility that the object knew exactly where he was hiding but had merely been toying with him. Regardless, for a time, this incident convinced Gorman that these intruders—human or otherwise—were not omnipotent.

EXTRATERRESTRIALS

The best-known, but not necessarily the most credible, model for multiple anomalous phenomena is that extraterrestrials from a distant planet have been, and still are, visiting planet Earth and for some unknown reason the ranch as well as the surrounding Uinta Basin for decades. But it is very difficult to conceive of an agenda for a group of extraterrestrial visitors choosing such a remote and out-of-the-way location. In the standard model of the ET hypothesis, the aliens are flying nuts-and-bolts spacecraft and interact with their surroundings in relatively understandable ways. The testability of the ET hypothesis lies in accumulating sufficiently robust and repeatable data. Overall, the events at the ranch yielded insufficient data to support or eliminate this hypothesis.

However, some of the events on or near the Utah ranch very much fit the standard description of UFOs. (We recognize that UFOs and ETs are in no way synonymous.) For example, within a couple of weeks of selling the property to NIDS, Tom and Ellen saw a silver-colored, disc-shaped object hovering over the ridge. The object, which was twenty to thirty feet in diameter and apparently metallic, brilliantly reflected the midday sun. As the Gormans drove onto the property, the object moved quickly from the ridge and then hovered directly over the ranch. Without warning, the silver disc disappeared in a bright flash of light. The Gormans felt unnaturally elated by the episode. This exhilaration appeared quite out of place given the extreme stress that the family had been under since arriving on the ranch. The silver disc they saw very much fits the classic description of an unidentified flying object as reported by thousands of people worldwide.

Another example of a "classic" UFO sighting occurred on November 13, 1996, at 1 A.M., when a colleague and I witnessed a silent, extremely fast-moving object coming from the north and rapidly executing a perfect loop over the command and control center before returning north. Again, the rapid speed, perfect maneuverability, and eerie silence fit the descriptions of a typical UFO, which the public tends to view as synonymous with extraterrestrials. But these sightings of "nuts-and-bolts" craft constitute a distinct minority of the reported events when measured against the total spectrum of weirdness on the property.

Could the various orbs have been unmanned extraterrestrial probes? Some of the more daring members of the SETI (Search for Extraterrestrial Intelligence) community have predicted that if intelligent life from somewhere else had begun to explore the cosmos, it would do so by launching small smart probes that could report back over the vast interstellar distances. This strategy would lessen the obvious costs of organic beings having to make the arduous exploratory trip across interstellar space. Thus, one prediction from University of Toronto professor emeritus Allen Tough is that if Earth is, or has been, visited by ET, we should see some evidence of such small, smart probes.

The ability of the blue orbs to negotiate skillfully through the branches of a tree on the ranch, to dexterously dodge the snarling jaws of dogs (often by mere inches), to inflict unspeakable fear on both Tom and Ellen Gorman all suggest an advanced technology and intelligence. Were the blue orbs examples of small smart probes?

NIDS was actually able to test a version of this "extraterrestrial probe" hypothesis. In the 1990s, aerospace engineer T. Roy Dutton published a predictive model based on thousands of global sightings of anomalous aerial phenomena over a period of fifty years. Dutton proposed that the data was most consistent with orbiting unmanned (presumably) extraterrestrial surveillance craft, which on a predictable basis released smaller probes that entered Earth's atmosphere. He believes that the entry into, or exit from, Earth's atmosphere of what he called small "scout ships" is responsible for the so-called UFOs seen by thousands of people around the world. What we found most interesting about Dutton's work was that he claimed to be able to specifically predict, from latitude and longitude coordinates anywhere on the planet, the timing of the appearance of such anomalous aerial phenomena.

Naturally, NIDS considered the Utah ranch a perfect location to test Dutton's predictions. But human and camera surveillance deployed specifically to monitor at the precise times predicted by Dutton (plus or minus twelve hours) failed to detect any evidence whatsoever of anomalous activity on the ranch or in the air space above.

NIDS also attempted to acquire real time spectra of the mysterious flying lights. We deployed an extremely efficient, custom made, portable spectrometer on scores of occasions in order to capture spectrographic data and came up empty-handed on this front as well.

NIDS participated in multiple expensive forays into the analyses of purported extraterrestrial artifacts and even alleged extraterrestrial biological tissue, but without exception the results of our analyses yielded only terrestrial signatures. In fairness, we should point out that the many attempts made in the past to analyze physical objects or artifacts said to come from UFOs yielded not a single piece of evidence of extraterrestrial origin either. All artifacts, pieces of metal, or other objects from purported anomalous origins have yielded uniformly terrestrial signatures after being subjected to metallurgical, chemical, physical, biological, or other analysis. This includes the analysis of isotope ratios in metals from unusual sources, all of which yielded nothing but terrestrial signatures. To our knowledge, not a single piece of metal or artifact acquired in nearly sixty years of research has survived scrutiny and is still regarded as genuinely anomalous.

ANCIENT ASTRONAUTS

A version of the extraterrestrial hypothesis holds that alien interest in Earth is not new, but that, in fact, colonizing ETs or gods have been on Earth, the moon, Mars, or somewhere close by for thousands of years and have routinely interacted with humans since time immemorial. This hypothesis holds that the ETs may be responsible for the Bible, for the ancient legends of the gods in multiple cultures, for building some of the ancient monuments, etc. This scenario may be consistent with the concept of angels and demons from the Bible and other religious beliefs. Whether these beings are angelic or demonlike ultimately depends on belief. Quite aside from such considerations, there is no physical data to support this version of the ET hypothesis either.

CHAPTER 26
The Military

Could the phenomena seen at the ranch be the result of the military's newest ultrasecret toys? If this hypothesis is true, then the Uinta Basin has been an outpost for the testing and deployment of military gadgetry and advanced technology since the advent of the military-industrial complex in the late 1940s. Such military technology may include advanced camouflage techniques; Special Forces insertion, extraction, and assassination techniques; psychological warfare techniques; advanced holographic technology and aircraft; drone and unmanned combat vehicles; and more.

What might the military's agenda be? To test advanced technology on unwary individuals in isolated rural communities who would be less likely to raise a fuss about being "experimented" on? Both northeastern Utah and northern New Mexico have a four- or five-decade history of intensive, sustained anomalous activity involving hundreds, maybe even thousands, of people subjected to bizarre, unexplained phenomena. These communities exist far off the main highway system. From personal experience interviewing scores of the residents, both communities fiercely protect their privacy and are not predisposed to reporting what they experience to either strangers or the media. Both areas are relatively closed populations with low standards of living and both contain Native American reservations. In this context, if one were going to conduct a program of testing new technological toys on an unwitting population without much fear of getting caught, these isolated rural groups might be ideal. Of course, this kind of program would be a violation of civil rights and thus would be illegal. Does the illegality rule out this possibility? We think probably not.

In their efforts to leave no stone unturned, NIDS personnel invited input from "remote viewers" concerning the activities at the Utah ranch. Remote viewing is a controversial but intriguing mental technique that has been investigated and developed in classified studies sponsored by military and intelligence agencies in the United States, Russia, China, and other countries. Although it is widely considered to be a form of psychic ability, proper remote viewing is conducted according to structured scientific protocols.

A gifted remote viewer named Ingo Swann provided the spark that led to a decades-long study of remote viewing sponsored by American intelligence agencies. The primary research was conducted by the Stanford Research Institute under the direction of physicist Hal Puthoff and was funded by the CIA. Puthoff and his team succeeded in documenting results that can rightfully be described as mind-boggling. In essence, remote viewers are seemingly able to project their consciousness across space and time and to access information that is unavailable by other means. Viewers are given a random series of numbers and letters that serve as target coordinates but that have no relation whatsoever to the target itself and are intended solely as a focal point for the viewers. All tests are blind, in that viewers are given little or no information about the target itself, although test managers sometimes give them questions about the kinds of information they should seek.

The CIA's remote-viewing program was eventually transferred to the Defense Intelligence Agency (DIA). In the late 1970s, the U.S. Army created an operational remote-viewing unit and used it to gather information about foreign adversaries. Overall control of the program was then turned over to the DIA in the mid-1980s and operated under the code name of "Star Gate." The program was transferred back to the CIA in 1995 and was then canceled after the public release of a controversial report that declared that remote viewing had little or no value as an intelligence tool. This conclusion was so inconsistent with the astounding results that have been documented over three decades that most people familiar with the capabilities of remote viewing speculate that the research program is ongoing, but as an even more secretive operation.

While we do not place undue weight on any results obtained from remote viewing, we also did not want to omit any possibilites. This research project, after all, was not following any beaten path. By the very nature of the subject matter, NIDS had to keep an open mind and be willing to explore all avenues. In that spirit, NIDS asked several capable remote viewers to view the ranch on four different occasions. One was a former government employee and is regarded as the most uncannily accurate remote viewer alive today. Although he had no information about the property and no knowledge of what had taken place, his off-the-cuff peek at the target produced a nearly exact sketch of the ranch, including the canal, the ridge, and the individual homesteads. He identified a spot on the southwest portion of the ranch as harboring an energy that he described as "disturbing."

A second remote viewer quickly homed in on the middle homestead as the center of what he described as weird energy vortices. That homestead was the location of much of the most intense activity recorded on the ranch.

Another prominent remote viewer, a retired military officer who was active in the CIA/DOD research program in its early stages, was asked to provide his impressions concerning the daylight mutilation of a calf. He said he had the sense that a robotic drone had carried out the mutilation, and that the drone might be of interdimensional origin. He then added that the drone had some connection to the military.

A fourth remote viewing exercise also suggested a military connection for the events at the ranch. Angela Thompson, a published author and trained remote viewer, teaches RV techniques to interested students in Boulder City, Nevada. At our request, she asked her associates in the Nevada Remote Viewing Group to take a look at the Gorman ranch, although none but Thompson were given any information about the target. Thompson's group consists of remote viewers around the country. They offer their services to corporate clients but also to humanitarian projects and law enforcement investigations.

Five remote viewers agreed to participate and were given a random coordinate, SR110202. They were told that the target is a location and were asked to describe events that occurred at the location in July 1996. They were later asked to describe current activity at the location, as well as activity five years hence. NIDS received a report on their findings in January 2003. The overall impression of the group was that the location was the site of a military operation, an operation that most likely involved the U.S. Navy. One viewer described the Pentagon. Two perceived images of a navy ship, perhaps an aircraft carrier, which was somehow linked to the activity in Utah. The viewers saw armed men in military uniforms who wore dark sunglasses and sported navy tattoos.

Remarkably, three of the five independently perceived a large grid, net, or honeycomb made up of sophisticated electronic equipment and wires that had been embedded in the ground. Most of the activity at the location, they said, was taking place outdoors and at night. One viewer mentioned sparkling lights in a dark sky, something akin to a Fourth of July display. Another perceived the presence of a humanoid with a long neck and large head who spoke a "foreign" language. Overall, they came away from the task with feelings of dread, nervousness, darkness, and death, which is consistent with the impressions of the previous remote viewers.

Though it sounds far-fetched, the idea that some sort of military facility might be located in the underground had surfaced before. Tom Gorman and his family said they often heard the sounds of heavy machinery or metal equipment coming from under the earth. The fact that the previous owners had warned them to avoid any digging is a curious footnote to this scenario. Why give such a warning?

We offer one additional anecdote in this regard. Following the appearance of a mysterious ice circle in a shallow pond, a local psychic who walked the property declared that the circle had been produced by a technology that was located underground. We do not ascribe any level of credibility (or noncredibility) to this psychic impression but mention it because of its similarity to the events listed above and because of its consistency to the stories and legends of the Ute tribe, whose members have long believed that some kind of presence dwells in the ground beneath the Gorman property.

We emphasize that the impressions of psychics and remote viewers do not constitute solid evidence of anything. NIDS simply hoped that the remote-viewing strategy could generate additional leads that we could investigate independently; we did *not* expect remote viewing to generate information that would stand on its own. We relate these investigations only in the interests of exploring all options and possibilities. The suggestion, psychic or otherwise, that the ranch was the site of some sort of clan-

destine military experiment cannot be ruled out entirely. The
military and intelligence communities have vast resources at
their disposal. They undoubtedly possess technological capabilities that are far beyond anything that is known in the public sec-
tor. And in the past, they have certainly demonstrated a
willingness to carry out psychological warfare exercises, disinformation campaigns, and surveillance programs against unwitting civilian populations, even when such programs are clearly
illegal.

Could the military be responsible for the phenomena observed at the ranch? The fact that military personnel were seen
on several different occasions in the vicinity strongly suggests
that the property became, at some point, a topic of interest for
intelligence operatives. Whether they were attempting to monitor an unknown phenomenon or were gauging reactions to their
own secretive endeavors is not known and may never be known.
The U.S. government isn't exactly cooperative in freedom of information matters these days. More material is being classified
than ever before, and military officials have been advised to
stall, delay, or deny requests for information, no matter how benign. However, we are reasonably confident that the military
scenario is not a satisfactory explanation for all or most of what
transpired.

Does the United States government possess technology that
could duplicate many of the events and phenomena that were
documented on the ranch? Holographic projections might conceivably account for some of the UFO sightings. We also know
of advanced camouflage experiments that could theoretically explain some of the unusual intrusions or poltergeist-type incidents. It is even conceivable, if unlikely, that a new generation
of silent, hovering Stealth-like craft has been developed and was
deployed in a remote corner of Utah for purposes that cannot be
discerned at this time. But if such technology is being used to
terrorize innocent Americans, to inflict permanent psychologi-

cal scars on entire families, to butcher animals and destroy the financial resources of law-abiding, taxpaying citizens, then such activity is not only illegal, it is despicable. No excuse, even the threat of global terrorism, could justify such an endeavor.

Our military probably wishes it could do some of the things that have been seen on the ranch, but it seems doubtful to us. Blue, glowing, glasslike, electrically charged orbs filled with swirling liquid, orbs that act and react intelligently, independently, and instantaneously, orbs that are capable of vaporizing dogs and of terrorizing people and animals, are well outside the realm of any known or projected military research programs. This statement is made with some confidence, based on our long-standing personal relationships with persons who are in positions to know such things.

Has any mind-control research yet perfected methods by which individuals can be made to feel intense fear or unexpected elation, as was reported on the ranch? Such a capability would undoubtedly come in handy, if it could ever be mastered by the United States government. Again, we have no way of answering this question, but we are skeptical it would be used on U.S. citizens if it existed.

Although holographs or other projected illusions could conceivably explain some UFO or creature sightings on the ranch, they do not explain the physical evidence that accompanied those sightings. Some of the creatures that were seen left footprints in the snow and the soil. The unknown animal that attacked a horse left bloody wounds along the legs of the horse. The entity that stripped forty-five pounds of flesh from a newborn calf in a matter of minutes during broad daylight certainly left behind some physical evidence. Some physical object, definitely not a holographic projection, created large, distinct impressions in the grass and dirt.

One final possibility should be mentioned from the perspective of military activity on the Utah ranch. Recent allegations have surfaced that the Air Force Office of Special Investigations (AFOSI) engaged in several deception and disinformation operations in the 1970s, the 1980s and (presumably) in the 1990s. Many of these operations involved the simulation of "UFOs," the manufacture of bogus evidence indicative of "extraterrestrial visitation" designed to conceal classified military technology or simply to lead investigators astray. In 2005, retired AFOSI special agent Richard Doty broke his silence to publicly acknowledge being involved in several of these "alien visitation" operations, the most famous being the disinformation campaign to persuade Albuquerque physicist Paul Bennewitz that an alien base existed in Dulce, New Mexico. The operation is described in detail in Greg Bishop's book *Project Beta: The Story of Paul Bennewitz, National Security, and the Creation of a Modern UFO Myth*.

Since Dulce appeared to be the site of AFOSI orchestrated bogus alien and UFO operations, was the Utah ranch also in the cross-hairs of another AFOSI (or other agency) disinformation campaign? Was the huge plethora of phenomena experienced by the Gormans and by NIDS simply a well-orchestrated deception that was designed to create yet another UFO or "alien" myth? While the motivation for such an exercise is unclear, it is difficult to refute, on a case by case basis, that many of the incidents that the Gormans and NIDS experienced could have been created by a creative and very skilled team of deception artists. But the motivation? A possible explanation for targeting the Gormans was that prior to the family's arrival, the abandoned property was being used for an important purpose and the operation was designed to drive the Gormans off the land. Or was the intent simply to create yet another "UFO" legend? Were the later ranch incidents designed to lead NIDS, a high credibility, well-funded organization that was spending considerable resources, astray? We will probably never know. Regardless, although we deem it unlikely, in the light of Richard Doty's allegations, we cannot completely refute the possibility that the Utah ranch was the target of a deliberate and very skilled deception operation.

But what technology could account for a living, breathing, bulletproof wolf, a beast that seemingly disappeared in midstride but left behind a chunk of its decaying flesh? What military or intelligence program can pull off a trick like that? What invisible soldier repeatedly slipped into Ellen Gorman's locked bathroom (see page 239) and removed her towel and hairbrush? What Delta Force commando infiltrated her kitchen and unpacked her groceries from the cabinets? How many covert operatives did it take to surreptitiously invade the Gorman home for the purpose of taking the spatula out of the frying pan so that it could be hidden in the freezer? Which tough-as-nails marine was assigned the vital but routine task of switching the salt into the pepper shaker and the pepper into the salt shaker? To put it mildly, it's a bit of a stretch.

As for the UFOs and anomalous aerial objects, it could be argued that our military might currently be capable of replicating some of the objects that have been seen on the ranch by the Gormans, by NIDS observers, and by others. But it is doubtful that any government on earth had such technology more than fifty years ago, which is when the sightings of huge, structured, metallic saucers and discs in the Uinta Basin became a matter of public discussion. Hundreds of well-documented sightings have been recorded in the area in the ensuing decades. What's more, the Ute tribe lived in the basin long before the CIA or U.S. Air Force existed, long before the United States itself existed.

The Utes say the unusual aerial objects have always been present. They also say that the Gorman ranch has been a repository of dark, evil energies for at least fifteen generations. The military did not invent these perceptions.

CHAPTER 27

The Native American Connection

ould the skinwalker curse somehow explain the various unusual phenomena that have been reported at the Gorman ranch over many decades? We are hesitant to endorse the existence of either skinwalkers or curses in any objective or literal sense, but there is no question that the story, as told by the Utes for a century or more, hangs over the ranch like a dark and ominous cloud. It is the umbrella explanation the Utes have embraced to try and make sense of otherwise inexplicable events.

While mainstream scientists are unlikely ever to give credence to any theory based on tribal lore or the black magic powers of shape-shifting Indian witches, it is difficult to ignore the seeming connection between the best-documented paranormal hot spots around the country and a strong Native American presence. Indigenous tribes seem to be on the fringes of nearly all of these paranormal outbreaks. Where you find one, you almost always find the other. The Uinta Basin is the most notable example, but there are several others, including Yakima, Washington, and Dulce, New Mexico, as we have already mentioned.

The San Luis Valley of Colorado is another location that fits the profile. It is the largest alpine valley in the world, eighty miles long, fifty miles wide at some points, with a floor that sits seventy-five hundred feet above sea level. Mount Blanca, the fourth highest peak in Colorado, dominates the skyline. More to the point, the San Luis Valley has long been the site of well-documented incidents of high strangeness. It is the place where the first publicized case of an animal mutilation occurred in 1967. Not coincidentally, the valley has also been the site of hundreds of UFO sightings over several decades and easily ranks as one of the most intense UFO hot spots on the planet.

Journalist Christopher O'Brien, who has lived in the San Luis Valley since 1989, has chronicled a rich tapestry of paranormal events in the vicinity, including continuing incidents of animal mutilation, frequent sightings of UFOs and mystery helicopters, and numerous eyewitness reports of Bigfoot encounters. He says the valley's paranormal legacy extends back centuries, and that one of the first Spanish explorers to enter the valley wrote diary accounts about weird flying lights in the sky and powerful humming noises that emanated from underground.

Not surprisingly, the region also oozes Native American mysticism and legend. The Yuma culture was in the valley five thousand years before the birth of Christ. The list of tribes, bands, and peoples that are known to have moved in and out since then is long. Among those indigenous groups that managed to survive into this century, the San Luis Valley is almost universally revered as a special, mystical place. The Tewa Indians, descended from the Pueblo people and now living in New Mexico, believe that the San Luis Valley is the equivalent of the Garden of Eden. The Tewas say the first humans to enter this world crawled up through a hole in the ground to escape their previous plane of existence. Native Americans who live in the valley today say they were taught that the Creator still lives in the mountains that surround San Luis and that He sometimes appears to humans in the form of a Sasquatch.

It is the beliefs of the Navajo, though, that are more pertinent to this book. Like many other tribes and bands, the Navajo visited, hunted in, and inhabited the San Luis Valley, off and on, for hundreds of years. Historians believe that the Navajo were finally ousted from the valley by none other than the Utes. It is a development the Navajo people are not likely to forget, since they regard the valley as a sacred place and a fundamental cornerstone of their culture. Mount Blanca, the fourteen-thousand-foot peak that towers over the valley, known to the Navajo as Tsisnaasjini', the Sacred Mountain of the East, is revered as one of the four mountains chosen by the Creator as a boundary for the Navajo world. It is considered to be an essential component in the Navajo quest to live in harmony and balance with both nature and the Creator. If the Navajo were Christians, Mount Blanca would be their Bethlehem. If they were Jewish, it might be their Wailing Wall.

At a minimum, the San Luis Valley provides another example of a place that has experienced an extraordinary litany of high strangeness events, a "paranormal Disneyland" in the words of Chris O'Brien, while also being of great significance to Native Americans, a place drenched in tribal mysticism. The intersection of these factors may be meaningless, or at most coincidental, but considering that there are several other examples involving this same unlikely confluence of unusual circumstances, it at least deserves to be noted.

Sedona, Arizona, is yet another example. Long before Sedona became an artsy Mecca for New Age believers of all stripes, it was hallowed ground for Native Americans. The long-gone Anasazi believed the area to be the center of the universe and the home of the gods. More recently, Sedona has been transformed into a haven for spiritualists, channelers, UFO enthusiasts, and assorted free spirits, drawn by the town's mystical vibe and by persistent stories about an energy vortex that just might be a portal to other worlds or realities.

Hard-core skepticism isn't Sedona's strong point, and it is prudent to carry more than a few grains of salt when evaluating extraordinary claims emanating from the locals. One case that seems to have merit is eerily similar to the events endured by the Gormans. Over a two-year period in the early 1990s, a ranching family named Bradshaw persevered through a frightening series of unusual events. Their tribulation began with frequent sightings of glowing orbs in the sky, then progressed to poltergeist events in their home, highly dramatic Sasquatch episodes, sightings of gray "aliens," brushes with some sort of invisible being, the mutilation and harassment of their livestock and dogs, and the appearance of a portal of light. The Bradshaws say they could see another world on the other side of the portal, a description remarkably similar to things seen on the Utah ranch.

In 1995, a book about the episode was published called *Merging Dimensions*. The writers, Linda Bradshaw and Tom Dongo, arrived at the conclusion that there are rips or openings in the fabric of reality, and that these openings can create merging points between different dimensions and different realities. The Bradshaws think that the entities and energies they encountered were from some other reality and that they were able to slip in and out of our world through the merging point or portal that had somehow opened on their ranch. The Bradshaws had no idea that during the same time period they were enduring their own series of encounters with the unknown just outside of Sedona, the Gormans were trying to cope with similarly bizarre activities in northeastern Utah.

Native American beliefs about alternate worlds pop up as well in a provocative work of fiction by the late, great Louis L'Amour, who wrote more than a hundred novels in his prodigious career. Readers bought more than 225 million copies of his books. Thirty of his novels were made into movies. L'Amour was best known for his western sagas, tales of gunfighters and lawmen, good guys on galloping horses and bad guys on the receiving end of frontier justice. But L'Amour's final novel, released a year before his death, represented an abrupt departure from what his readers had come to expect.

The title of the book was *The Haunted Mesa* and its focus was Indians, not cowboys. Relying on a lifetime of dogged research and personal experience, L'Amour plunged headfirst into the topic of Native American mysticism and spiritual beliefs. The result was a book that closely parallels some of the possibilities that are central to the mysteries of Skinwalker Ranch.

L'Amour based *The Haunted Mesa* on his understanding of Navajo and Hopi beliefs concerning "other worlds." Hopi spiritual leaders teach that our current earthly plane of existence is the fourth world the tribe has known. They traveled into this fourth reality, they say, by passing through a door or tunnel that opened in their previous world. The Hopi view closely parallels that of the Navajo. The Navajo are taught that they entered this world via a tunnel in the earth and that they departed their previous reality in order to escape from an unspecified evil. In his narrative, author L'Amour relies on his principal characters to argue for the existence of other realities. The story line implies that the other worlds from which the Hopi and Navajo escaped are most likely other dimensions, and that the doors between these dimensions are sometimes traversable. It's a fictional account, of course, but is firmly based on tribal religious traditions, which, from one perspective, seem to be in sync with prominent theories now being championed by cutting-edge physicists. In essence, Native Americans have believed for hundreds of years in the existence of such concepts as parallel universes, alternate dimensions, and traversable wormholes, although this isn't the terminology used by the tribes. To think they arrived at their beliefs without the benefit of Ivy League educations, particle accelerators, or Doppler-based calculations is certainly curious.

Though we have no empirical evidence to prove that a Navajo skinwalker might really have the black magic ability to put a curse on the Utes, thus triggering a century or more of weird activity near the Ute reservation in Fort Duchesne, there is plenty of historical evidence to suggest the Utes certainly had done enough to deserve a curse or two, at least from the Navajo perspective. The Utes were often at war with the Navajo, and those wars were fought in places of interest to this book.

The Utes' traditional homeland encompassed most of the state of Colorado, including the San Luis Valley and Elbert County, both of which, as we've noted, are areas of high strangeness. In the San Luis Valley, the warlike Utes drove the Navajo out. In the early 1800s, the Utes allied themselves with the Jicarilla Apaches (the same tribe that now occupies the paranormal hot spot of Dulce) in a bloody war against the Navajo, all at the behest of the Spanish. The Utes considered much of present-day New Mexico and Arizona to be their homeland. It could be argued that the Utes have historical ties to nearly all of the areas that are now regarded as paranormal hot spots. And there is no question that they, along with most of the other tribes in the American West and Southwest, consider the subject of witchcraft to be very serious business.

According to Tom Gorman, the Fort Duchesne Utes were not happy to learn that NIDS had purchased the Gorman ranch. But NIDS received gracious cooperation and friendly openness from the tribe. Some tribal members initially warned Gorman that the NIDS purchase was a big mistake. In hindsight, it appears they were worried that a team of scientists poking around might annoy or arouse whatever malevolent force might inhabit the place. Tom Gorman came to believe that some of the Utes were dabbling in black magic rituals themselves. One day, he discovered the carcass of a raccoon that had been mutilated in ritualistic fashion and was left splayed on top of a tree stump in a manner so obvious that he believed he was meant to find it. Someone had slipped onto the property to perform what Gorman suspected was some sort of black magic ceremony. Was it a warning of sorts?

The title of this book mentions a hunt for skinwalkers, so it's fair to ask whether we found any. We have no real evidence to suggest that Indian witches have the ability to pull off the kind of wild exhibitions and phantasmagorical events that were seen on the ranch by many witnesses. Native American beliefs in other worlds and portals to other realities have demonstrable equivalents in modern science, and in that sense, we wondered if their legends about skinwalkers might also have a scientific stepbrother, something that might help us to understand what unfolded.

It may not have been a true skinwalker that was haunting the Gorman ranch, but something certainly manifested itself, again and again, over a period of several years. *Something* gave rise to the bewildering plethora of anomalies. Something manifested itself as a black cloud in the trees, from which a telepathic voice reached out to the NIDS scientists. Something appeared as an opaque "Predator" creature that roared loud enough to make a grown man cry. Something moved through water but could not be seen. Something generated voices from the sky. Something floated in the sky as glowing orbs, Stealth-like craft, disc-shaped UFOs, and even a flying recreational vehicle. We saw them. Either a single "intelligence" was responsible for conjuring up this amazing variety of ephemeral events, changing its appearance and form as well as its tactics and strategies, or a "family" of anomalies was sharing the rent in some kind of paranormal fraternity house. Skinwalker or not, it could be argued that the ranch, for unknown reasons, functioned as a one-stop supermarket for all manner of bizarre activity.

Though we can't say with any certainty whether the things that happened at the ranch represent the presence of one entity or force, or whether there was more than one "intelligence" at work there, we can speculate about intentions. Was "the skin-walker," for lack of a better term, a malevolent force? In a sense, it was. It butchered and abducted livestock, attacked other animals, and seemed intent on generating strong emotional responses from the Gormans, as if it thrived on terror and confusion.

But it also seemed to draw a distinction between harming animals and harming humans. Other than psychological damage, intense headaches, nosebleeds, and a few unexplained cuts, no humans were physically maimed, while animals certainly were. Whatever might have been slipping into our world seemed to know the difference between slicing up a newborn calf and doing the same to a human. It stuffed bulls into a trailer, but didn't do the same to people. Dogs were incinerated, but the Gormans were not. If we are to believe both the Navajo and academic researchers, skinwalkers are inherently evil and would not hesitate to kill a person. It suggests that whatever was present on the ranch was not a skinwalker in any traditional sense.

CHAPTER 28

Other Dimensions

D ress up the skinwalker hypothesis in the garb of modern physics and the question becomes: Could the phenomena at the Utah ranch have been the result of interdimensional beings or time travelers accessing our space-time continuum, or physical beings jumping across vast interstellar distances via wormholes or stargates? It might sound like science fiction, but attempts to relate current physics theories to observed anomalous phenomena are being presented at conferences and published in scientific journals by physicists such as Eric Davis, Hal Puthoff, Bernard Haisch, Michio Kaku, Beatriz Gato-Rivera, Jack Sarfatti, and others. A burgeoning physics literature, including papers from mainstream science luminaries such as Kip Thorne and Matt Visser, have described the possibility that wormholes are shortcuts through the cosmic neighborhood, thus circumventing the velocity of light limitation for interstellar travel. Visser, Davis, and others have published the mathematical requirements for traversable wormholes as well as approaches toward engineering wormholes in mainstream physics journals.

According to some variations of the interdimensional hypothesis, the creatures or beings that access our physical reality need not necessarily be physical but may be capable of influencing our physical reality and/or may be capable of mind control. They may manifest their influence either in long-term evolutionary or cultural manipulation of human affairs, or they may have even more inscrutable agendas. Variations on this theme include Jacques Vallee's "control system hypothesis," Patrick Harpur's "daimonic reality," and Michael Grosso's "imaginal realm," as well as John Keels's "ultraterrestrials" and the older Middle Eastern concept of the djinn.

It's worth noting that Harpur's book *Daimonic Reality* describes in vivid detail several of the strange animals seen on the Utah ranch, including the large cat and huge doglike creatures shot by Gorman on March 12, 1997, the bushy-tailed brownish animal seen by the Gormans in April 1999, as well as some of the weird "semi-physical" attributes of these creatures. Harpur says these creatures come from a "daimonic" (meaning "alternate" and not to be confused with *demonic*) reality and are characterized by both their physical and ghostlike attributes. The wide spectrum of events witnessed is not inconsistent with Harpur's description of a crossover of beings and creatures from another world into ours. The confusing array of unpredictable occurrences, the strange animals that appear to exist in some ambiguous world that is neither 100 percent physical nor 100 percent immaterial, and the tricksterlike quality of this apparently precognitive sentient intelligence, all fit a pattern of activity that has existed on Earth for millennia, according to Harpur.

Prior to the arrival of NIDS, the Gormans were tormented by tricksterlike phenomena almost daily. The trickster activity was a constant irritation and the backdrop against which all of the other more spectacular events occurred. For example, in the midst of calving season, Ellen Gorman purchased several boxes of cereal so that the family could eat quickly on the run. But when it came time to eat the cereal, no one could locate the boxes. They were later found in the fridge, freezer, and oven. But that's not all. Irrigation headgates were found inappropriately opened or closed several times with no tire tracks or footprints in the vicinity. Water hoses would disappear and then be found in unusual places, always rolled into a neat circle three to four feet in diameter. In numerous instances, the shovels for digging irrigation ditches could not be located when needed but were found in unusual places later. And several times, when Ellen Gorman was getting into the shower, she would place her towel and hairbrush on the counter near the shower. But when she got out of the shower, they were gone. The items would then turn up in odd places in other parts of the house.

And there was more still. Doors in the home would suddenly open and close with great force for no apparent reason. On one occasion Ellen Gorman had gone grocery shopping and had carefully put all the items on the kitchen table before stacking them in the cupboards. The job took a considerable amount of time. But when she returned several hours later, all the items were back on the kitchen table where she had originally placed them. This particular incident upset Ellen considerably because Tom suggested that maybe she had just forgotten to put the items away. But Ellen clearly remembered doing so.

The Gormans' world was anything but black and white. The family got so used to finding the salt in the pepper shaker and vice versa that they would always shake a small amount onto their hand as a test prior to putting it on their food. On one occasion, Tad and a couple of his friends had been asked by Tom to move 150 to 200 large metal corral poles from the front yard to a location beyond the canal. The poles were of varying length and were seven to eight inches in diameter and weighed between 25 and 150 pounds each. It took the three teenagers about four hours of hard work to complete the job. The boys finished about noon.

Tom returned around four in the afternoon to find all the poles back almost precisely where they had originally been in the front yard. He asked Tad why they had not done the job yet. The boys expressed considerable frustration over this incident, since they had spent several hours doing just what they had been told to do. Interestingly, the poles were replaced almost, but not quite, in the same spot as they had originally been, since the original depressions left by the heavy weight were visible on the ground beside the newly replaced poles. The family endured this constant, frustrating, unnerving trickster activity for almost two years. According to Harpur, this sort of activity is diagnostic of intrusions from other realities, which have been an ongoing part of humanity's experience for millennia.

After experiencing all sorts of bizarre activity during his research on UFOs, John Keel formulated his "ultraterrestrial" hypothesis, which postulates that Earth has shared living space for millennia with other intelligent beings who interact with humans when they choose to, who are more intelligent than us, and who manipulate our physical and psychic reality for their own obscure agendas. "Within a year after I had launched my full time UFO investigating efforts in 1966," Keel writes in *Operation Trojan Horse*, "the phenomenon had zeroed in on me, just as it had done with the British newspaper editor Arthur Shuttlewood, and so many others. My telephone ran amok at first, with mysterious strangers calling day and night to deliver bizarre messages 'from the space people.' Then I was catapulted into the dream-like fantasy world of demonology. I kept rendezvous with black Cadillacs on Long Island and when I tried to pursue them they would disappear impossibly on dead-end roads . . . Luminous aerial objects seemed to follow me around like faithful dogs. The objects seemed to know where I was going and where

I had been. I would check into a motel at random only to find that someone had made a reservation in my name. I was plagued by impossible coincidences, and some of my closest friends began to report strange experiences of their own—poltergeists erupted in their apartments, ugly smells of hydrogen sulphide haunted them . . ."

Keel's descriptions of existing in the strange netherworld between reality and some deeply disturbing nightmare exactly encapsulate the Gormans' description of what life was like on the ranch prior to NIDS investigation. Of course, when NIDS arrived, this tricksterlike activity diminished significantly. Occasionally, however, one of the NIDS researchers would wake up in the middle of the night with an incredibly strong sulfurous odor emanating from one corner of the room. It was invariably the same NIDS researcher who experienced this. NIDS researchers also experienced the constant tricksterlike interference with the dog runs in May 1997, and perhaps the mysterious destruction of the three video surveillance cameras could be categorized as tricksterlike. Other than these few exceptions, life for the NIDS staff was mercifully free of the sort of constant irritation the Gormans experienced with the trickster.

How likely is the interdimensional hypothesis? Unfortunately, we don't know enough about the world for this hypothesis to stand on its own two feet—there is no known experiment, for example, that will distinguish between the simple ET hypothesis and the interdimensional hypothesis without more robust data. From a scientific perspective, extraterrestrial and the interdimensional beings are indistinguishable from one another—they both come from other worlds.

CHAPTER 29

Outer Worlds

Let's take a step back and examine the broader picture by considering a question about perspective posed by the physicist Michio Kaku. "Let's say that a ten-lane superhighway is being built next to an anthill," he says. "The question is: would the ants even know what a ten-lane superhighway is, or what it's used for, or how to communicate with the workers who are just feet away? And the answer is no . . . If there is [another] civilization in our backyard, in the Milky Way galaxy, would we even know its presence? . . . There's a good chance that we, like ants in the anthill, would not understand or be able to make sense of a ten-lane superhighway next door."

Other thinkers have posed similar questions. If a pair of pigeons alighted on a discarded newspaper page on a New York street, would they understand the content of the paper? A herd of cattle, grazing in a pasture, would be unlikely to appreciate the physical beauty of the surroundings, let alone be able to comprehend the prospects of their pending trip to the slaughterhouse.

Spanish physicist Beatriz Gato-Rivera has questioned whether we could be immersed in a larger civilization without being aware of it. She notes that "typical civilizations of typical galaxies" would likely be hundreds of thousands, or even millions of years more evolved than our own. She compares it to a family of mountain gorillas and asks if the gorillas could possibly know that they are "a protected species inhabiting a natural reserve in a country inside the African continent of planet Earth," blissfully unaware of nations, borders, religion, or politics, or of their own position within the planetary pecking order.

"Would any country on this planet send an official delegation to the mountain gorilla territory to introduce themselves openly and officially to the gorilla authorities?" Gato-Rivera asks. "Would they shake hands, make agreements, and exchange signatures with the dominant males?" She argues that it is reasonable to conclude that, just as gorillas could not possibly grasp the intricacies of the larger world around them, we humans may simply lack the brainpower to comprehend the most basic foundations of a larger reality that surrounds us. Gato-Rivera and other physicists suspect that we are already a part of a much larger civilization, whether we can fathom this or not. And if the multiverse paradigm is true, there could be untold numbers of older, infinitely more advanced civilizations that might be capable of traveling into our world at will.

"If there exist thousands, or millions, of parallel universes, separated from ours through extra-dimensions," Gato-Rivera suggests, "it would be natural then to expect that some proportion of these universes would have the same laws of physics as ours . . . , and many of the corresponding advanced civilizations would master the techniques to travel or 'jump' through . . . the extra dimensions. This opens up enormous possibilities."

Physicists at the University of California at Davis have theorized that our known universe might exist within some sort of giant black hole that is "reversed," a so-called outward, a universe that is "expanding into a much bigger place." According to mathematical physicist Blake Temple, the implication of this is that "there is something outside. Our universe is in a much larger space-time. The universe we know could be much bigger than the regions of expanding galaxies."

But the question is, could someone "out there" in a larger reality or parallel world get here and interact with our lives, in subtle, overt, or incomprehensible ways? It is possible, physicists say, but it is questionable whether we would even notice, and even more doubtful that we would understand it if we did notice.

"Aliens may be here now, in another dimension, a millimeter away from our own," Kaku suggests. He points out that wave frequencies from alternate realities, other universes, and other times are all around us every moment of every day. "However, just like you can only tune into one radio channel, you can only tune into one reality channel, and that is the channel that you exist in. The catch is that we cannot communicate with them, we cannot enter these universes."

Still, he and some of his colleagues agree with the premise that other, more advanced civilizations may have the technological ability to make the jump at will. Kaku even speculates that we will be able to make the same jump ourselves, someday far in the future. Trillions of years from now, when all of the known stars burn themselves out, our universe will undergo what might be called a "big freeze." It will be impossible for any intelligent life to survive. When that happens, we will have no choice but to depart for warmer pastures.

University of Pennsylvania physicist Max Tegmark, the most outspoken proponent of the multiverse concept, believes that parallel universes already interact in definitive ways with our humble cosmic abode. Tegmark and other theorists think that our universe once had nine dimensions, but during the early stages of cosmic expansion, the three dimensions that constitute the parameters of our world stayed put, while the other six dimensions grabbed a cosmic taxi and split. Space may be nine dimensional, but matter may exist only in a three-dimensional surface, also known as a membrane, or brane, for short. Tegmark cites the work of renowned physicists Paul Steinhardt of Princeton and Neil Turok of Cambridge University, whose calculations indicate the existence of a second three-dimensional brane that is literally parallel to our own plane of existence but separate from us because it is just a tiny fraction higher on the dimensional scale. It's there, and it is part of our current reality, whether we can see it or not.

"This parallel universe is not really a separate universe because it interacts with ours," Tegmark says. This interaction presumably is ongoing and omnipresent, even as you read this sentence.

If this single universe of ours truly is infinite, if countless other universes exist in our little slice of reality, if parallel dimensions also exist, each with its own countless universes, if it is likely that our world is part of a much larger reality, and if it is possible, according to the laws of physics, for an advanced intelligence somehow to travel between those alternate realities, then it may also be possible to begin to understand the kinds of mysterious events that have been observed and reported by humans throughout recorded history, including those that have been documented on the Utah ranch.

Distinguished scientists and prominent journals are no longer reluctant to discuss the most exotic theories about the nature of reality, so long as those discussions restrict themselves to the realm of esoteric chitchats between fellow highbrows. But when scientific outsiders or maverick thinkers raise the possibility that other realities might be manifesting themselves in our physical world right now, most scientists retreat behind the snooty barricades of academia and rationalism. They may heartily support the mathematical equations that establish the existence of the other worlds but recoil in intellectual revulsion whenever so-called paranormal events are reported, no matter how compelling the physical and testimonial evidence might be.

The late astronomer J. Allen Hynek of Northwestern University, who served as the principal debunker in Project Blue Book, the much-maligned U.S. Air Force investigation of the UFO phenomenon, told the story about a convention of fellow astronomers that he attended. According to Hynek, one scientist who had slipped out of the convention for a smoke witnessed a UFO in the sky above the gathering. He ran back into the convention and announced with some alarm to the crowd that a flying saucer was hovering outside. Not one of the assembled scientists got up even to take a peek. To do so would have given credence to a discredited and unsubstantiated superstition and would have invited the scorn of their esteemed colleagues. Flying saucers *can't* exist, therefore they *don't*.

At its most basic level, science is supposed to represent the investigation of the unexplained, not the explanation of the uninvestigated. Yet few scientists are willing to risk the criticism of their peers (or the withdrawal of their research grants) if it means pursuing the subjects that are deemed by unofficial acclamation to be unworthy tabloid fodder, the rants of disturbed minds, or the folklore of drunken trailer park lowlifes. It can't be, therefore it isn't.

The ideas of visionary thinker Giordano Bruno were so unsettling to the political, religious, and scientific establishment of his day that he was burned at the stake for espousing them. The modern science establishment, which is viewed by some as the equivalent of a harsh and unforgiving religion with its own strict commandments and rigid code of conduct, no longer burns its outcasts, but it certainly excommunicates those who stray too far from the fold. Unusual experiences that intrude into our daily lives are routinely discarded or excluded from any serious discussion through the time-tested techniques of ridicule, trivialization, and debunking. A witness who reports a UFO is either mistaken, psychotic, inebriated, or out to make a buck.

It is no small irony that millions of people all over the world have experienced and reported events of high strangeness, events that are completely consistent with the most exotic, cutting-edge theories of the nature of reality, yet these encounters are almost always dismissed without so much as a cursory investigation. Despite nearly a century of mind-boggling, paradigm-shifting discoveries and ideas to the contrary, we continue to live as if Newton's laws still define our reality, as if the high strangeness of the universe has nothing to do with our daily lives. But is this true?

In 1991, a nationwide Roper poll found that millions of Americans were experiencing events that defy our textbooks and our understanding of reality. The study polled nearly six thousand people and was easily the largest statistical sample ever undertaken concerning "paranormal" experiences among the general public. The size of the sample meant that the statistical margin of error was 2 percent or less, far more reliable than most political surveys. Astoundingly, the Roper pollsters found that 18 percent of those polled had experienced waking paralysis (with no apparent cause); 13 percent said that they had experienced "missing time," hours or even days for which they have no memory or explanation; 8 percent said that they had seen balls of light or orbs, which might be interpreted as UFOs; and another 8 percent noticed unexplained scars on their bodies.

The survey suggests that approximately one in every ten Americans, more than twenty million people, had experienced events that might be considered paranormal. A second Roper poll conducted seven years later found a smaller number of people, 7 percent instead of the earlier 10 percent, reporting these same experiences. It isn't clear why the overall percentage changed, but even at the lower number, millions of people claimed to have experienced something strange and dramatic in their lives.

It is clear that mainstream public attitudes and beliefs toward so-called anomalous phenomena are profoundly out of step with mainstream science attitudes and beliefs. And just as cosmologists and physicists sometimes sound like religious mystics, an increasing number of psychologists and anthropologists are beginning to explore the human mind as something more than the consequence of neurochemical trafficking in the brain. These new areas of research may also be relevant to an understanding of some of the events that occurred on the ranch.

CHAPTER 30

Inner Worlds

Does the wide variety of anomalies observed at the ranch have any connection with human consciousness? Of course, much of mainstream science simply dismisses such reported anomalies as products of *pathologies* of human consciousness. But what if they are instead *doorways* from human consciousness into other realities? People such as writer Patrick Harpur, philosopher Michael Grosso, psychologist Kenneth Ring, among others, argue that there are separate realities inhabited by other intelligences and that humans can sometimes access these places. The central question is whether these separate realities are real, external, and physical in any sense, and whether they somehow tie in to human consciousness.

Certain events that took place at the ranch suggest a possible link among shamanism, human consciousness, and whatever force was operating there. The Gormans became convinced that the phenomena eavesdropped on them and often reacted to what they said. One example was the "transfer" of four two-thousand-

pound bulls into a tiny trailer. This event happened almost immediately after Ellen Gorman's private conversation with her husband in which she expressed extreme apprehension at the idea of
losing the animals. Was this sheer coincidence? If not, how could
their conversation have been overheard in a vehicle on a remote
ranch? After scores of similar coincidences, the Gormans ended
up believing that the intelligence knew what they were thinking
and could even anticipate what they were going to do.

It is tempting, therefore, to speculate that a link existed not
only with the Gormans' physical presence on the ranch but also
with their mental states. A similar link apparently took place
with a team of scientists at Lawrence Livermore Laboratory in
California during the 1970s when they were testing the famous
Israeli psychic Uri Geller. The scientists, all of whom had top secret security clearances, began to experience all manner of
weird phenomena, both in their homes and at work shortly after
they began to study Geller. Suddenly, gray flying saucers would
materialize inside their homes or labs. Huge birds would turn up
at the end of their beds, strange metallic voices would speak to
them by telephone, and other objects would become visible and
float in the air in front of them. It was as if some sort of doorway to another world had been opened for them. Only after they
stopped studying Geller did their lives return to normal. Were
these strange occurrences diagnostic of some kind of "disturbance" in the field of consciousness?

There is some evidence to suggest that forms of meditation
(and therefore altered states) were connected to the appearance
of anomalous phenomena on the ranch. The "Predator incident"
of early 1996 appears to have been provoked by an individual's
conscious attempt at meditation. And a little over a year later,
less than sixty minutes after an hour-long meditation by an investigator, a creature crawled through a tunnel of light in midair
at exactly the same spot where the meditation had taken place.

But human consciousness at the opposite end of the spectrum from the stillness of meditation apparently also provoked anomalous events. Just prior to the purchase of the ranch by NIDS, Tom Gorman says that an incoherent woman arrived one day in 1996. As the woman got out of her car and began talking to Tom in the front yard of the homestead, a nearby tree began to shake violently and the leaves began to rustle loudly despite a total lack of wind. Suddenly the woman, who admitted to being mentally disturbed, began to scream loudly and pointed at the tree. She described the presence of legions of demons and monsters in the shaking tree. Tom couldn't see the "demons," but he could plainly see the whole tree shaking. After ushering the crazed woman off the property, the tree returned to its previous stillness. Tom thought that somehow the woman's unhinged mental field had disturbed something in the environment and provoked the "psychic" outburst in the tree.

Eventually, after being exposed to scores of bizarre but unrepeated events, some members of the NIDS Science Advisory Board began to hypothesize that a sentient, precognitive, nonhuman intelligence occupied the ranch. One board member even suggested that this intelligence was engaged in deliberately provoking emotional reactions in the residents, perhaps in order to utilize those emotions in some way. This notion is not so farfetched given the Gorman family's long held suspicions that something or someone was watching them constantly and was aware of their every thought and movement. The cat-and-mouse mind games experienced by the NIDS researchers, the trickster-like interference with the dog runs in the summer of 1997, and the chilling destruction of our surveillance cameras in 1998, an event that was captured on videotape, all lent support to the notion that some kind of manipulative intelligence was present. While it is admittedly difficult to establish the existence of such a precognitive, nonhuman intelligence, the thought was never far from the minds of those of us who walked those fields, day and night, for so many years.

One of the more celebrated proponents of alternate realities and their link to states of human consciousness was John Mack, a Harvard University professor of psychiatry and author of two widely read books on the phenomenon of alien abduction. Mack studied some two hundred cases of alleged alien abductions over a ten-year period before a drunk driver in London killed him in October 2004. Over time, Mack had come to believe that many of the events reported by persons who claimed to have been subjected to so-called alien abductions actually occurred in other realms or dimensions that humans had somehow accessed, though not at a normal level of consciousness. Mack saw overlaps between the alternate realities occupied by shamans and the world from which these "aliens" came to terrorize their "victims" during alien abductions.

The experiences that Mack's patients reported were visceral, vivid, and powerfully disturbing. On many occasions they were life changing. The visitors emerged from another world or else pulled the victim into a separate reality where they performed sometimes very painful examination procedures reportedly involving, at times, the collection of bodily fluids. In many instances, the victims were forcibly examined without their consent; a situation tantamount to kidnapping. Mack believed that these powerful traumatic experiences, if eventually accepted by the victims, were the doorways for spiritual evolution. Many abductees say they received instruction from their abductors about the systematic destruction of Earth and the need for ecological awareness, a sort of "alien boot camp" concerning the necessity for taking better care of the natural world. Mack eventually concluded that the abduction experiences were not happening physically but were associated with some other level, dimension, or reality that was either a part of human physical reality or could easily influence it.

The "inner world" nature of these experiences is further

strengthened by the recent work of psychiatrist Rick Strassman. In one of the few scientific studies of its kind, Strassman studied the effects of injecting the hallucinogen N-N-dimethyltryptamine (DMT) into sixty volunteers at the University of New Mexico between 1990 and 1995. The research was funded by the National Institutes of Health and followed extremely rigorous research protocols and human subject methodologies. Strassman was startled to discover that a large number of his subjects experienced meetings with "alien beings" and nonhuman intelligences following injections of DMT. The details of these meetings were remarkably consistent among the subjects, though none of these experimental volunteers were allowed to compare notes prior to being interviewed.

To Strassman, the almost identical "hallucinations" experienced by different subjects in their encounters with the "aliens" seem to imply that the subjects experienced some kind of objective reality as a result of the DMT injections. He hypothesized that perhaps DMT was able to open a perceptual doorway into other "dimensions" where the human mind could meet and communicate with the denizens of these realities. In Strassman's words, "I suggest DMT alters the receiving qualities of the brain, and employ a television analogy. Personal healing occurs by an enhancement of 'contrast and focus'; invisible worlds and entity contact takes place by changing reception of 'channels' to include dark matter and parallel universes." Strassman points out that the human brain makes DMT naturally. This leads him to speculate that perhaps, under certain circumstances, in certain states, the human brain might increase its levels of endogenous DMT and so make it easier for the "alien abduction" experience to take place.

Or could an external agency manipulate the DMT levels in the human brain in order to make journeys to other worlds possible, or alternatively, such that the denizens of other worlds could cross over into ours? Or was some unknown environmental variable on the Utah ranch, perhaps very low frequency electromagnetic radiation, responsible for triggering DMT cascades in the brain? Was such a mechanism responsible for some of the incidents? It would have been interesting to conduct regular and comprehensive blood chemistry monitoring of all the ranch residents and investigators. Did they have elevated levels of DMT? Were glucocorticoid stress hormones elevated? Although we discussed performing such blood tests, they were never carried out.

Shamans, who apparently have the ability to cross over into other worlds at will, also report having DMT-like experiences. Michael Harner, who is arguably the father of Western shamanism, learned to be a shaman while living among the Jivaro Indians in the Ecuadorian Andes and the Conibo Indians in the Peruvian Amazon. He had a very interesting, and perhaps relevant, vision of an "alien intelligence" that was directly intertwined with human consciousness. It happened in the early stages of his exploration into shamanism.

"Now I was virtually certain I was about to die," he noted after entering an ayahuasca-induced trance. (Ayahuasca is a hallucinogenic drink made from the bark of a South American vine; it contains several psychoactive ingredients, including DMT.) "As I tried to accept my fate, an even lower part of my brain began to transmit more visions and information. I was 'told' that this new material was being presented to me because I was dying and therefore 'safe' to receive these revelations. These were secrets reserved for the dying and the dead, I was informed. I could only dimly perceive the givers of these thoughts: giant reptilian creatures reposed sluggishly at the lowermost depths of the back of my brain, where it met the top of my spinal column. I could only vaguely see them in what seemed to be gloomy, dark depths. Then they projected a visual scene in front of me. First they showed me the planet earth as it was eons ago, before there was any life on it. I saw an ocean, a barren land and a bright blue sky. Then black specks dropped from the sky by the hundreds and landed in front of me. They flopped down, utterly exhausted from their trip, resting for eons. They explained to me in a kind of thought language that they were fleeing from something out in space. They had come to planet earth to escape their enemy."

Harner continued: "The creatures then showed me how they created life on the planet in order to hide within the multitudinous forms and thus disguise their presence. Before me, the magnificence of plant and animal creation and speciation—hundreds of millions of years of activity—took place on a scale and with a vividness impossible to describe. I learned that the dragon-like creatures were thus inside all forms of life, including man. They were the true masters of humanity and the entire planet, they told me. We humans were but receptacles and servants of these creatures. For this reason, they could speak to me from within myself."

Harner's encounter might link to Jacques Vallee's hypothesis that the phenomenon represents a technologically advanced control system that may reside on this planet and that the control system is responsible for the displays of the UFO. This control system seems to interact with humans on multiple levels, from the nuts-and-bolts level all the way through to the psychic level. Vallee hypothesizes that inner changes as well as long-term societal transformations that are inspired by the assorted encounters with this control system are far more significant than the physical trappings of the encounters themselves. Though the control system seems to operate for reasons that are opaque, mysterious, and largely unfathomable by human beings, its "productions" may have an agenda to "educate" human societies over an extended period of time. The appearance of the Virgin Mary at Fatima and at Lourdes might be examples of this control system's productions. The question is: Were the events at the Utah ranch another one of their "productions"?

There are, in fact, some profound shamanic overtones to what the Gormans experienced. The parallels to a story told by a

Siberian-born medical doctor named Olga Kharitidi are quite obvious. Kharitidi spent many years practicing psychiatry in Novosibirsk, then a technological center in Siberia. While accompanying a friend of hers who was seeking a healer, Kharitidi found herself in the company of a shaman from the Siberian mountains of Altai. During her friend's encounter with the shaman, Kharitidi witnessed events that defied her understanding of physical reality. As a trained psychiatrist in the Western medical model, her first impression was that she might be experiencing a psychotic episode as she watched the shaman perform "impossible" feats. She says that it was not uncommon to see UFOs or other unusual aerial phenomena appear in the sky just as the shaman was beginning to "access other worlds." The shamans in Kazakhstan supposedly pay little attention to these UFO-like manifestations. They seem to regard them as distractions from their mission of journeying to and contacting "other worlds." In fact, the shaman explained to Kharitidi that the UFOs were the equivalent of wallpaper or window dressing around the entrances to these other worlds and should not be the focus of the effort.

Most of the possible links between consciousness and anomalous phenomena remain at the anecdotal level, and our investigation of the events at the Utah ranch did not dramatically change that. Were denizens from another dimension playing a cat-and-mouse game with the Gormans and us? Or were these denizens who haunted the ranch simply skilled in the art of camouflage? Was an unknown force from an "inner world" attempting to manipulate consciousness? While our efforts and experiments to answer these questions did not succeed in any definitive sense, it wasn't for a lack of trying.

CHAPTER 31

Revolutionary Science

In February 2002, as most of the paranormal activity at the Gorman ranch slowed to a mere trickle, NIDS scientists received a calling card of sorts, a little reminder that weird things could still happen. A perfect circle appeared overnight in a shallow, ice-covered pond not far from the main house.

The circle was carved into the ice, approximately one-quarter inch deep and just under six feet in diameter. It was mystifying how this could have been done, let alone why. The ice was so thin that any sort of weight on top would have broken it. There were no footprints in the mud that surrounded the pond, except hoof prints from the cattle that grazed nearby.

The NIDS team reacted as scientists should. Close examination of the circle indicated that the counterclockwise motion of a sharp object had delicately carved a circular impression in the ice and had caused ice shavings to accumulate at the edges of the grooves. The investigators collected the ice shavings from the circle and some from a control spot elsewhere on the ice and placed them into sterile test tubes for later study. All shavings from the ice circle as well as controls were subjected to elemental analysis by X-ray fluorescence to look for any residues, possibly metallic, from whatever sharp instrument had so delicately carved out the ice circle. The tests yielded nothing of note. The scientists also took readings for magnetic fields, electrical fields, and radiation in a hundred-yard radius around the ice circle. They checked the cattle and the surrounding environment for any unusual markings or tracks but found none.

Ice circles have periodically been found in other parts of the world. In Russia, much larger circles have been seen and photographed on frozen lakes. Witnesses claim that the deep impressions were caused by UFO-like craft that were seen touching down. As far as we know, though, there are few precedents for the small, fragile carving that appeared in the ice pond on the ranch, although a short time later a family from Connecticut sent NIDS a photograph of a perfectly drawn twenty-foot-diameter circle in the ice outside their home. However the ice circle on the ranch was accomplished, it required a deft touch to make the carving without breaking ice so thin. And for what purpose? Was it meant as a teasing reminder that while NIDS might be close to giving up on the ranch for the lack of significant activity, the intelligence behind it all was still around, still unseen, and still a mystery?

Four centuries ago, the predecessors of modern scientists were struggling to move beyond the "magic" and "magical thinking" that were inherent in alchemy. Since the age of enlightenment, there has been an ever-increasing gap between the scientific method and mainstream science's attitude toward anomalies. Today's science establishment conveniently forgets that Sir Isaac Newton, one of the icons of modern scientific thought, spent far more of his time studying and writing about alchemy than he did about physics.

When faced with the mystery of the ice circle, the NIDS scientists reacted as they were trained and in full compliance with the rules of modern science. They took samples, searched for clues, and analyzed what physical evidence was available. Yet it is likely their colleagues would excoriate them for giving even a moment's thought to something so weird and seemingly insignificant. Scientists today generally shun anomalies. The weirder the anomaly, the greater the professional stigma against studying it. Nevertheless, the study of anomalies is central to the discovery process in science. Studying anomalies can open important doorways through which science can enter while traveling toward revolutionary discoveries.

There is a distinction between "normal science" and "revolutionary science." In his controversial book, *The Structure of Scientific Revolutions,* Thomas Kuhn argued that the majority of scientists engage in normal science. They stand on the shoulders of giants from the previous generations and rarely make giant leaps forward. Kuhn writes, "Normal science does not aim at novelties of fact or theory, and when successful, finds none. New and unsuspected phenomena are, however, repeatedly uncovered by scientific research and radical new theories have again and again been invented by scientists . . . Discovery commences with awareness of anomaly, i.e., with the recognition that nature has somehow violated the paradigm-induced expectations that govern normal science." Of course, Kuhn was referring to experimental anomalies, not necessarily the physical anomalies found on the Gorman ranch, but his message still applies.

The lessons that Kuhn drew for the experimental anomalies that pop up during the practice of "normal science" are equally well suited to science's first hesitant attempts to study the physical anomalies that abound in the "paranormal." The latter anomalies are simply more extreme versions of what Kuhn had in mind.

According to Kuhn, there are a small number of revolutionary scientists. These often lonely individuals or groups are engaged in creating paradigm shifts or in the creation of new scientific disciplines. They are the solitary people on the road less traveled who often have to endure the hostility or marginalization by colleagues. But, according to Kuhn, the scientists engaged in revolutionary science are the ones who make a difference. They make possible the giant leaps forward. They are the visionaries.

As might be expected, Kuhn's book, when it was first published, led to a firestorm of criticism because it insulted the hard work of the majority of scientists who are engaged in "normal" science. And, of course, most scientists prefer to think of themselves as the harbingers of new paradigms. But the majority of scientists are, according to Kuhn, mere technicians, elaborating upon, or tinkering with, the ideas of those who have gone before them. Kuhn called this process of normal science "puzzle solving."

Any investigation of paranormal events would certainly be categorized as revolutionary science. After all, such phenomena do not have any precedent according to scientific studies and superficially appear to break the laws of science. When a scientist begins approaching these extreme anomalies, there are huge risks, not only to his or her career but also to the work itself, since it necessitates the adaptation of scientific methods to a topic where predictions, theory, and previous experimental evidence are either absent, scanty, or incomplete. It is in this strange land that the scientist must learn to navigate while at the same time try to persuade his or her colleagues that such a study has inherent value and is not a complete waste of time and effort. In approaching something that has no precedent, great care must be exercised against gullibility while at the same time maintaining an open mind. For many scientists, walking this fine line is impossible.

Yet we argue that the study of extreme anomalies can follow the same rules that Kuhn laid out for experimental anomalies. The study of extreme anomalies has great value in accelerating the discovery process. The scientific method is built on precedence, repeatability of experiments, and having enough data to make testable predictions. When a phenomenon under study refuses to obey these rather narrow strictures, what happens? What happens when a possibly intelligent phenomenon refuses to be predictable? Does a scientist walk away? Should the NIDS team have simply pretended that the ice circle never appeared, or that the calf dismemberment was a figment of their imagination?

Is it possible to utilize the scientific method to study an intelligent, extreme anomaly? Herein lies an enormous challenge that is made doubly difficult if the majority of one's scientific colleagues do not even accept the reality of the phenomenon under study. Publication of data in scientific journals becomes difficult. Presentation of data at scientific conferences becomes impossible except for those marginalized societies that are devoted to studying these phenomena. And to put it bluntly, these societies are poorly attended, have almost no funding, and exist at the boundaries of the scientific community. In other words, they have no impact or influence on science or on how it is conducted.

Obtaining grants to study or research these phenomena is impossible without publications in mainstream science journals, which in turn becomes impossible if there are insufficient resources to gather the data. These factors that accumulate in the sociology of science make the task of initiating the study of anomalies extremely difficult.

Political perceptions and peer pressure contribute to the reticence of scientists to tackle unpopular topics. In March 2005, the Associated Press reported the results of a study conducted by researchers at three major universities. The study found that many, if not most, scientists and researchers shun controversy when choosing research topics. They avoid topics that might be frowned upon by colleagues, concerned that their professional reputations might suffer. More than half of the scientists interviewed said they felt constrained by informal and unspoken rules about what should and should not be studied. In addition, of course, the federal government, which controls the bulk of all science research dollars, frowns upon controversial research projects. Federal funds for research into paranormal topics are all but nonexistent. Small wonder then that so many modern scientists choose to stick with "safe" research projects and goals.

The advent of NIDS with comparatively large resources due to the philanthropic generosity of real estate and aerospace entrepreneur Robert Bigelow aimed to at least remove the lack of resources from this part of the equation. However, even with substantial resources, scientists at NIDS still had to face the daunting task of studying something for which very few previous hard data existed. In fact, there was little or no hard data even to suggest the existence of these phenomena. Whatever reliable information existed was scattered across multiple, generally low-quality science journals. The challenge faced by the NIDS scientists was to come up with some methodologies that might address these meager precedents. In doing so, it was very necessary to walk that intangible line between gullibility and overdone skepticism that might result in missing the boat altogether.

In approaching the study of extreme anomalies, it has become possible to recognize that there is essentially no difference between the gullible believers and the extreme skeptics. Although superficially both camps seem poles apart and frequently attack each other, in reality both groups are the same in many respects. Both have abandoned the cautious attitude of the scientist and both have stopped thinking critically. Both are in fact equally useless, as they tend to impede, and even sabotage, the study of extreme anomalies. Both the true believer and the knee-jerk skeptic have the effect of muddying the water and of obscuring the fragile data in a deluge of noise. Both extreme points of view contribute to the reluctance of mainstream scientists to plunge into the study of anything that is even remotely tainted by the paranormal label.

The penalty for violating the unwritten prohibition against investigating forbidden topics can be harsh, even for well-established professionals. John Mack, a brilliant Harvard psychiatrist and Pulitzer Prize winner, was nearly run out of academia when his interest in alleged alien abductions became common knowledge. Mack was not only subjected to a fifteen-month-long inquisition by university lawyers and skeptical colleagues that threatened to strip him of his tenure and his job, he also became the butt of jokes and the object of ridicule among contemporaries who simply could not fathom how someone of his stature could suggest that "reality" may not be what it seems. He was accused of being a UFO nut and true believer, a scientist who had abandoned the sacred protocols of his profession.

"It's often said that I'm a believer and have sort of lost my objectivity. I really object to that," Mack told a TV interviewer, "because this is not about believing anything. I didn't believe anything when I started, I don't really believe anything now. I'm come to where I've come to clinically. In other words, I worked with people over hundreds and hundreds of hours and have done as careful a job as I could to listen, to sift out, to consider alternative explanations. And none have come forward."

Mack survived his inquisition, with considerable assistance from Harvard Law professor Alan Dershowitz, and he continued his research until his untimely demise. Instead of pulling back from controversy, he plunged forward into more controversial territory. He came to believe that there was some sort of connection among all manner of so-called paranormal activity, including telepathy, remote viewing, psychokinesis, near-death experiences, spirit manifestations, crop circles, shamanic journeys, UFOs, the power of prayer, and much more. Mack came to advocate the ultimate blasphemy—a new concept of reality.

"Taken together, these phenomena tell us many things about ourselves and the universe that challenge the dominant materialist paradigm," he wrote. "They reveal that our understanding of reality is extremely limited, the cosmos is more mysterious than we have imagined, there are other intelligences all about (some of which seem to be able to reach us), consciousness itself may be the primary creative force in the universe, and our knowledge of the physical properties of the physical world is far from complete. The emerging picture is a cosmos that is an interconnected harmonic web, vibrating with creativity and intelligence, in which separateness is an illusion."

Mack came to believe that although the scientific method is valuable, even essential, for studying those phenomena that make up the material world as we understand it, the rigid structure of modern science simply can't cut the mustard when it comes to evaluating other realities or those phenomena that seem to straddle the known world and "unseen realms." In contrast, the NIDS team resolutely stuck to the tenets of the scientific method. NIDS conducted an excursion into what Jacques Vallee elegantly described as "forbidden science," but the scientific method remained intact even though the experimental terrain became surreal.

The ranch in Utah offered the possibility of a laboratory of unusual happenings, none of which were easily classifiable. It is easy to understand why science has always stayed away from studying these subjects. How does one begin? Why should these subjects be studied? We argue that one answer is that these anomalies do indeed offer a window into studying potentially new models of physical reality, as well as offering new directions in physics and psychology. And studying anomalies offers scientists the potential to build a bridge into new areas and disciplines where science can operate. No phenomenon really breaks the laws of science. A new phenomenon may create a disturbance in the status quo, it may open doors toward establishing new theories, and in doing so it may force scientists who are willing to venture into uncharted territory to break truly new ground.

For about four hundred years there has been a divide between the physical world and the "metaphysical world." The former is the province of scientists and the latter is the domain of theologians and mystics. Tradition has it that never the twain shall meet. But is this really the future for humanity? Science can continue only so long to ignore phenomena that literally invade people's lives and that have fundamental effects on the worldview of millions of people around the world.

The gap between the direct life experience of many people and the reality that is recognized by science is leading many to adopt an "antiscience" attitude. In the past thirty years, a growing segment of society has begun to view science and scientists with distrust and suspicion. Scientists are increasingly associated with the development of unbridled technology and a lack of ethics or morals. The discovery and deployment of the atomic bomb, of genetically engineered organisms, of biological weaponry are all laid at the feet of scientists and science.

The development of these technologies, some of which the public regards as having spiraled out of control, is seen as a reflection of hubris and arrogance in science. And at the same time, science ridicules and trivializes a lot of profound, but anomalous, experiences. People cannot help but wonder at the truth capacity of science if it completely denies the reality of a large number of their own experiences. Public opinion polls show that science and scientists are increasingly out of step with the people's worldview.

Though the benefits of conducting revolutionary science are obvious, it has frequently been argued, often persuasively, that the events on the Utah ranch did not necessarily lend themselves to classical hypothesis-driven scientific methodologies. A more appropriate methodology, given the phenomena we faced, might have been one utilized by the intelligence agencies (which in turn is based on scientific principles). This intelligence approach to the problem was explained a few decades ago by Jacques Vallee in his classic work *Messengers of Deception*. In that book, Vallee introduces a "Major Murphy" who enunciates some of the basic principles and approaches that the NIDS staff agreed should be followed more closely in the Utah ranch investigations. It is worth quoting the initial conversation between Vallee and Major Murphy at length.

> Then he posed a question that, obvious as it seems, had not really occurred to me: "What makes you think that UFOs are a scientific problem?"
>
> I replied with something to the effect that a problem was only scientific in the way it was approached, but he would have none of that, and he began lecturing me. First, he said, science had certain rules. For example, it has to assume that the phenomenon it is observing is natural in origin rather than artificial and possibly biased. Now, the UFO phenomenon could be controlled by alien beings. "If it is," added the Major, "then the study of it doesn't belong in science. It belongs in Intelligence." *Meaning counterespionage.* And that, he pointed out, was his domain.
>
> "Now, in the field of counterespionage, the rules are completely different." He drew a simple diagram in my notebook. "You are a scientist. In science there is no concept of the 'price' of information. Suppose I gave you 95 per cent of the data concerning a phenomenon. You're happy because you know 95 per cent of the phenomenon. Not so in Intelligence. If I get 95 per cent of the data, I

know this is the 'cheap' part of the information. I still need the other 5 per cent, but I will have to pay a much higher price to get it. You see, Hitler had 95 per cent of the information about the landing in Normandy. But he had *the wrong 95 percent!*"

"Are you saying that the UFO data we use to compile statistics and to find patterns with computers are useless?" I asked. "Might we be spinning our magnetic tapes endlessly discovering spurious laws?"

"It all depends on how the team on the *other side* thinks. If they know what they're doing, there will be so many cutouts between you and them that you won't have the slightest chance of tracing your way to the truth. Not by following up sightings and throwing them into a computer. They will keep feeding you the information they want you to process. What is the only source of data about the UFO phenomenon? It is the UFOs themselves!"

Some things were beginning to make a lot of sense. "If you're right, what can I do? It seems that research on the phenomenon is hopeless, then. I might as well dump my computer into a river."

"Not necessarily, but you should try a different approach. First you should work entirely outside of the organized UFO groups; they are infiltrated by the same official agencies they are trying to influence, and they propagate any rumor anyone wants to have circulated. In Intelligence circles, people like that are historical necessities. We call them 'useful idiots.' When you've worked long enough for Uncle Sam, you know he is involved in a lot of strange things. The data these groups get are biased at the source, but they play a useful role.

"Second, you should look for the irrational, the bizarre, the elements that do not fit . . . Have you ever felt that you were getting close to something that didn't seem to fit any rational pattern, yet gave you a strong impression that it was significant?"

The events that occurred on the Utah ranch certainly gave us the impression that they were significant. So Major Murphy was perhaps correct. This research project was beyond a simple scientific problem that was amenable to standard hypothesis-driven science. It involved hunting a very wily quarry. And NIDS constantly had to accept the possibility that any information acquired in this hunt was only the information that the intelligence (assuming that we were in fact dealing with an intelligence) wanted us to have. The phenomenon became much more elusive when NIDS took over the property in August 1996. Tom Gorman believed that the phenomenon took extraordinary measures to become much more opaque and hidden almost immediately after NIDS assumed control.

Our attempt to target a wily and deliberately evasive research subject is perhaps unprecedented in scientific research but is the norm in the cat-and-mouse games of espionage and counterespionage. For example, hunting the skinwalker transcended what wildlife scientists normally do to hunt or track wild animals because the target of the NIDS hunt proved over and over its capacity to keep a couple of steps ahead of us. According to Major Murphy, the art of intelligence gathering and the techniques of counterespionage are much more appropriate when dealing with active deception.

A second main ingredient in Major Murphy's advice suggests that NIDS should follow a more proactive, and less reactive, stance toward the phenomenon. One example of our proactive strategy had its genesis in a nugget of information Tom Gorman acquired during his protracted interactions with the phenomenon. He found that whenever he made changes to the topography on the ranch, for example by removing a tree line, or by digging a new irrigation ditch, there would be fresh appearances of mysterious flying objects. So NIDS tried several times to provoke the phenomenon by digging. In several instances neighbors reported sightings of a low-flying orange object that disturbed the animals within forty-eight hours after we had dug large trenches in symmetrical rows. But in no case did the surveillance cameras that constantly recorded the airspace above and on the ranch pick up the flying object. Nor were NIDS personnel able to provide visual corroboration of the object even though the object was said to be heading toward the Gorman property.

The NIDS team decided to take this proactive approach a step farther by attempting to initiate a direct dialogue with the phenomenon. Ideas were kicked around about possible ways to communicate with "the entity" or how to encourage it to communicate with us. In one experiment, we placed see-through Perspex boxes containing pictorial and alphabetic patterns at various points around the property. The premise, as slim as it might seem, was that the unknown intelligence might try to communicate directly with us by manipulating the letters or pictures in the boxes. Unfortunately, nothing happened. In retrospect, the experiment might seem far-fetched and overly hopeful, but we were navigating uncharted waters, looking for out-of-the-box ideas. Experimentation, after all, is supposed to be a fundamental component of the scientific method.

The investigation of the phenomena at the Gorman ranch was an ambitious if unconventional example of what science is supposed to do. Explore the unknown. Ask questions about the unexplained. Poke around and see what happens. Honest inquiry into unanswered questions is—or should be—a textbook definition of what science does.

But finding answers is not always part of that definition even when engaged in "normal" science. Making sense of the big picture is a tall order. Though we can eliminate a few of the hypotheses—hoax, group hallucination, and tectonic strain theory—there is simply insufficient data to be able to select a likely solution to the events among the remaining possibilities. Part of the reason for this is the incredible variety of paranormal experiences we encountered at the Skinwalker Ranch. That was one of the most unsettling aspects of our investigation. It's as if some cosmic puppet master had written a laundry list of every spooky phenomenon of modern times and then unleashed them all in a single location, resulting in a supernatural smorgasbord that no one could possibly believe, even less understand. The events were random and unpredictable, and never happened more than once in the same place or in the same way.

If there is an intended message or lesson in all of this, what could it possibly be? Needless to say, everyone who played a part in the investigation has logged many a sleepless night while pondering this central question without arriving at a satisfactory answer.

Today, the types of events recorded in this book are still occurring around the United States and in other parts of the world. They remain unexplained. And science continues to look the other way.

Epilogue

More than seventy years ago, the brilliant physicists Werner Heisenberg and Niels Bohr were involved in emotional shouting matches about how best to reconcile the "impossible" world being described by the equations of fledgling quantum mechanics with the world that we all live in. Since then, the physical models and mathematical constructs that are published in peer-reviewed journals have only become stranger, more bizarre, and almost frightening. They describe a universe that few of us would ever experience without having a nervous breakdown.

Traversable wormholes, parallel universes, and extraspatial dimensions all seem fascinating to us while we are sitting in our armchairs watching PBS or the Discovery Channel. These TV programs can be assimilated relatively comfortably, and then we go to bed, wake up in the morning, and go to work. The ideas, we think, have little or no bearing on the "real world." They are merely abstract physics concepts and have no real effect on our lives.

But suppose they intruded into our lives? Suppose that, against our will, we were plunged into a world where traversable wormholes were staring us in the face from only one hundred feet away. Suppose that, as happened to the Gorman family, we saw a completely different sky from another world on the other side of that wormhole? Suppose we came up against bizarre creatures and monstrous denizens of the "underworld" as they walked freely around our property? How would we react if the concepts of extra dimensions and parallel universes were interfering with our daily lives in a frightening and completely unpredictable manner? What then? Suppose we were suddenly living in a reality where we were certain that some nonhuman intelligence was aware of our every word and move and seemed to have a fascination with toying with us. What if these intelligences began introducing terrifying animals onto our property from other dimensions and, conversely, what if they stole our cattle and transported them into another dimension, never to be seen again?

This may have actually happened to a ranching family in remote Utah and, under the radar, possibly happened dozens of times around the United States to other ranching families. These families did not go in search of the bizarre, nightmarish realities they encountered. They are normal citizens who are concerned with paying the mortgage and spending time with their children.

Mainstream physics journals now describe time travel, macroscopic extradimensional spaces, and zero-point energy as serious topics. They are not part of some realm of fantasy. Rather, they appear to be legitimate ways of describing the real world, admittedly a world that few of us get to see. "Throughout mankind's cultural history," notes Hal Puthoff of the Institute of Advanced Studies at Austin and arguably one of the preeminent theoretical physicists in the world, "there has existed the metaphysical concept that man and cosmos are interconnected by a ubiquitous, all-pervasive sea of energy that undergirds, and is manifest in, all phenomena. This pre-scientific concept of a cosmic energy goes by many names in many traditions, such as ch'i, ki or qi (Taoism), prana (yoga), mana (Kahuna), brakah (Sufi), élan vital (Bergsonian metaphysics), and so forth . . . Contemporary physics similarly posits an all-pervasive energetic field called quantum vacuum energy, or zero-point energy, a random, ambient fluctuating energy that exists in so-called empty space." According to Puthoff, the sea of energy described from personal experience by some mystics is the same zero-point energy field described by the mathematical equations of breakthrough physics.

But again we must confront the question: Is this realm of existence to be accessed intellectually only by the arcane equations of breakthrough physics or by the single-minded self-discipline of the mystic? Is 99.9 percent of mankind shut out from experiencing the full dimensions of physical reality? If the events described in this book have any merit, the answer is obviously no. It is ironic that an apparently unbridgeable void exists between the strange reality experienced on the ranch and the realities of multiple parallel universes and dimensions and traversable wormholes that are read by physicists in mainstream physics journals every month.

Our collective perception of reality isn't what it used to be. The spectacular and relentless march of scientific progress has resulted in dramatic changes in our perception of the world around us. Slightly more than a century ago, scientists believed they were close to a full explanation and description of the universe. The principal building block of the universe, they believed then, was a mysterious substance known as ether. As theoretical physicist Stephen Hawking notes, the ether theory quickly collapsed, science moved forward, and "the world has changed far more in the last one hundred years than in any previous century."

Humanity has experienced dramatic upheavals in its perceptions of reality several times in the past, but these earlier paradigm shifts are mere pebbles trickling down a hill compared to the tectonic blockbusters that loom just ahead. Mankind is on the cusp of a fundamental, mind-blowing, all-encompassing change, a revolution that could dwarf all previous transformations.

Early humans, scratching for food and scrambling for survival, looked at the stars and had no idea what those twinkling lights in the sky might be. The first civilizations deified the stars and ascribed to them the personalities and powers of gods. Centuries later, Greek philosophers changed this worldview when they determined that stars are heavenly bodies whose movements circled Earth, still perceived then as the center of the universe. That prevailing paradigm shifted dramatically again when, a few hundred years later, Copernicus dared to suggest that Earth revolved around the sun, and that our sun was but one of many. The advent of the printing press helped usher in this Copernican paradigm, although the eventual shift in the collective perception was long and bloody.

Our current paradigm has led to a realization that there are millions of other galaxies in the vastness of the known universe. The larger the universe gets, it seems, the smaller and less important we humans appear. As one novelist put it, the history of astronomy is the history of increasing humiliation. Now the inferiority complex of our species may be due for another jolt, one that could change our view of reality far more profoundly than the previous paradigm shifts.

Think of how much of our view of reality has changed in our own lifetimes. Just a few years ago, no planets were known to exist outside of our own solar system. Scientists had reason to believe they might be out there somewhere but had no confirmation. By 2004, there are more than one hundred known extrasolar planets. Astronomers now accept that the planetary model of our solar system is likely the norm, which means, conservatively, there could be one hundred billion stars with planetary systems of their own in our galaxy alone. Observations from the Hubble telescope have shown that some of these planetary systems could be thirteen billion years old, almost three times as old as our little neck of the interstellar woods. "Innumerable suns exist," wrote Giordano Bruno in 1584. "Innumerable earths revolve around these suns in a manner similar to the way the seven planets revolve around our sun. Living beings inhabit these worlds."

Of the untold number of planets out there, how many might be suitable for the development of life? Not long ago, scientists believed that ours was likely the only world in this solar system to have ever supported life. By one current estimate, there could be two billion so-called Goldilocks planets, worlds that are neither too hot nor too cold and that, presumably, could support life as we know it.

Mainstream scientists generally acknowledge that the discovery and confirmation of extraterrestrial life would be among the most profound developments in human history. It would once again dramatically change how we look at the world and at ourselves. But a discovery of another sort could usher in a paradigm shift of even greater impact, so mind-boggling, so world shattering that none of us would ever be the same. The general public doesn't know it, but this shift, far subtler than a sudden discovery of ETs, is already under way. Cutting-edge scientists now accept its basic premises, although its profound significance is a long way from acceptance or even general understanding by the world at large.

It is hard enough for nonscientists to envision a single universe as vast as our own, one that was seemingly created out of nothingness in a single big bang some fourteen billion years ago. But what happens to our collective view of reality when the word filters down to the rest of us that our universe is only one of many? Today, the prevailing view among quantum physicists is that there is an infinite number of other universes, and that the structure of these universes may be far more exotic than we can fathom, involving parallel dimensions that are almost beyond the comprehension of our best minds. This concept is known as the multiverse or many worlds theory, and it has gained widespread acceptance in scientific circles.

"Is there a copy of you reading this article," asks physicist Max Tegmark in a recent issue of *Scientific American*, "a person who is not you but who lives on a planet called Earth, with misty mountains, fertile fields, and sprawling cities, in a solar system with eight other planets. The life of this person has been identical to yours in every respect. But perhaps he or she now decides to put down this article without finishing it, while you read on. The idea of such an alter ego seems strange and implausible, but it looks as if we will just have to live with it because it is supported by astronomical observations."

Tegmark and other leading physicists have now concluded that there are an infinite number of other worlds, other universes, other versions of each of us, living out lives that may vary from our own perceived existence by only the slightest detail. The roots of this theory have been around nearly eighty years. In the 1920s, physicists began to try and unravel the mystifying weirdness of quantum theory, which seemed to make sense in explaining the behavior of atoms but made little sense in explaining the visible world of people, cars, buildings, and other solid physical objects. Since that time, the world has carried on as if Isaac Newton's laws are still in effect, as if Einstein's relativity is the only way to explain reality. But the cold, hard truth of mathematical computations combined with rigorous observation of how things really work in the universe have slowly, gradually forced science to re-evaluate the fundamental nature of reality itself.

The first step toward reconciling the profound differences

between the microscopic world of quantum theory and the macroscopic world of relativity came in 1957 when a Princeton graduate student named Hugh Everett proposed that atoms and objects can, in fact, be in more than one place at one time. This was a major leap toward an explanation of the many worlds version of reality. It set into motion a thirty-year paradigm shift in which physicists gradually came to accept that our universe is a far stranger place than what we all learned in our school textbooks. By the 1990s, prominent articles in the mainstream press and in respected journals such as *Physics Letters, Physical Review,* and others openly addressed the fantastic possibilities of the multiverse paradigm. Scientists recognized that 90 percent of the physical universe is essentially invisible to any instruments we can devise. The fundamental inability of scientists to measure or predict the position and momentum of any elementary particle is clearly at odds with what science is supposed to do. A new theory was clearly needed.

Terms like *dark matter* and *dark energy* entered the scientific vocabulary in recent years. If 90 percent of the matter in the universe cannot be seen or detected, scientists asked, where is it? In response to this simple question, a concept known as string theory has gained gradual acceptance. According to string theory, the way to reconcile the existence of so much dark matter is to consider that it exists in parallel dimensions or alternate realities, invisible or undetectable by us. Depending on the version of string theory, there exist either eleven dimensions or twenty-six dimensions. Our own world consists of three known dimensions, or a fourth if you count time. The more exotic string theory version of reality did not immediately win over the skeptical world of science, but by 1999, an informal survey of leading physicists found that a majority now favors the multiverse concept.

"In other words," says science writer Marcus Chown, "physicists are increasingly accepting the idea that there are infinite realities stacked together like the pages of a never-ending book. So there are infinite versions of you, living out infinite different lives in infinite parallel realities." Obviously, the significance of this sea change in scientific thought hasn't come close to percolating down to most of the rest of us.

"The quantum theory of parallel universes is not some troublesome, optional interpretation, emerging from arcane theoretical considerations," writes Oxford physicist David Deutsch in his book *The Fabric of Reality*. "It is the explanation—the only one that is tenable—of a remarkable and counter-intuitive reality."

The idea is widespread. "We physicists no longer believe in a Universe," says Michio Kaku. "We physicists believe in a Multiverse that resembles the boiling of water. Water boils when tiny particles, or bubbles, form, which then begin to rapidly expand. If our Universe is a bubble in boiling water, then perhaps Big Bangs happen all the time."

Max Tegmark concurs: "The concept of the multiverse is grounded in well-tested theories such as relativity and quantum mechanics. It fulfills both the basic criteria of empirical science: it makes predictions and it can be falsified. Scientists have discussed as many as four distinct types of parallel universes. The key question is not whether the multiverse exists but rather how many levels it has."

Obviously, this is heady stuff and will not easily be assimilated into the general understanding of the public at large. Imagine, then, some of the more exotic possibilities of this theorem. If the universe is truly infinite, and an infinite number of other realities consisting of parallel dimensions and alternate universes are real, then everything that could happen *is* happening somewhere. Scientists speculate that there are universes where the laws of physics as we know them do not operate. Some universes must teem with life. Others must be completely dead. There are universes where time runs backward. The people there go to bed, then work backward through their day, taking off their pajamas and backwardly putting their suits and ties back on, walking in reverse back to their dinner tables, shuffling back to their cars for the reverse drive from home to their workplace, and ultimately back to the moment when their alarm clock went off that morning. In such a world, clocks not only run backward, but water glasses that are accidentally dropped and shattered on a kitchen floor miraculously reassemble in the hands of the person who dropped them.

In the same vein, alternate versions of history are played out in these parallel worlds. Somewhere, Hitler conquered the planet and buried the story of the Holocaust; Abraham Lincoln was killed while splitting logs and never became president; Bill Gates gave up on computers and drifted into a crack cocaine habit that landed him in prison; dinosaurs survived the impact of a killer comet, evolved into intelligent beings, and are now kicking back in their BarcaLoungers as they slam a few Reptile-Lite beers. Somewhere, a team of monkeys with typewriters has written the great American novel, along with a string of hit sitcoms.

Have you ever wondered what your life would be like if you had gone to a different school, accepted a different job, married a different person? In the multiverse, all of those other things have happened or are happening, and each of those individual alternatives has millions of its own parallel realities, different from the others by the nearly immeasurable factor of a single atom or quantum. Somewhere, a version of you is enmeshed in a completely different reality for the simple reason that the "you" in that other world had a ham sandwich for lunch instead of tuna, or because you had the flu on the fateful day when crazed eighth graders sprayed your junior high school with automatic weapons. Somewhere, as implausible as it might seem, a version of you is the CEO of Time Warner and is married to Carmen Electra. Lucky you. Or lucky other you.

Proponents of the multiverse theory do not emphasize such

fanciful scenarios when defending their idea. It's tough enough to champion a drastic revision of our fundamental understanding of reality without slipping off into what seems like the realm of science fiction. But this is precisely the point. The looming paradigm shift these scientists advocate is so fundamentally weird, disconcerting, and unbelievable that its acceptance by the larger public will take decades, or, based on previous experience, centuries. It isn't a subject that can easily be explained over dinner or even in a semester of physics lectures. Nonetheless, it appears to be true.

"Space appears to be infinite in size," writes Max Tegmark. "If so, then somewhere, everything that is possible becomes real, no matter how improbable it is."

What does any of this have to do with the events at the ranch in Utah? After all, millions of us have watched highbrow science shows on the Discovery Channel or PBS while stretched out on our living room couches. We've heard the sound bites from prominent thinkers as they spoke of parallel universes, extra-spatial dimensions, and traversable wormholes. We've rented the movie *Contact* and rooted for Jodie Foster as she overcame jealous rivals, religious zealots, and budget-minded bureaucrats in her pursuit of the ultimate truth about the structure of reality. And we are comfortable in our assimilation of these entertainments. We watch them, then click off the tube, go to bed, get up in the morning and drive to work, as if none of this has anything to do with our real lives. Abstract physics concepts might be a pleasant diversion on a Tuesday night, but they sure don't pay the bills on Wednesday.

But what if these fantastic scenarios suddenly intruded into our lives in direct, unmistakable, and frightening ways? How would we react if we were forcibly dragged out of our psychic comfort zone into a world where wormholes were staring us in the face from only one hundred feet away? What if those wormholes revealed to us an alien sky from another world, an incomprehensible glimpse of an alternate reality, as we stood in shock in our own front yard? What if bizarre creatures, long-extinct prehistoric beasts, and futuristic flying machines somehow seeped into our mundane existence from some other place and systematically assaulted our loved ones, our possessions, and our most basic concepts of reality?

Add to this seemingly improbable set of circumstances the presence of a pervasive, sometimes malevolent, nonhuman intelligence, an invisible trickster who knows our every thought, anticipates our every move, and seems intent on toying with us, terrifying us, and screwing with our daily lives, a presence that orchestrates a relentless, perverse, and unpredictable campaign of all-out psychological warfare, a campaign that ultimately makes us question our own sanity. Imagine the grotesque mutilation of prized livestock, the brutal incineration of family pets, frightening intrusions by disembodied voices and shadowy figures into the sanctity of a family's home, and inexplicable manifestations by unknown beings that cannot be harmed by guns or bullets. That's clearly what happened at the ranch in Utah.

What if it happened to you?

References

CHAPTER 2. LEGACY

Frank B. Salisbury, *The Utah UFO Display: A Biologist's Report* (Old Greenwich, CT: Devin-Adair, 1974).

CHAPTER 3. THE BASIN

Craig Fuller, "Uinta Basin," *Utah History Encyclopedia*, 2004, http://historytogo.utah/gov.uintabasin.html.

Robert Foster, "Buffalo Soldiers in the Utah Territory," *Wild West Magazine*, February 2000.

Louis Diggs, "The Buffalo Soldiers," 1995, www.louisdiggs.com/buffalo/.

Thomas Fouch, *Hydrocarbon and Mineral Resources of the Uinta Basin* (Salt Lake City: Publishers Press, 1992), p. 271.

CHAPTER 5. THE CURSE

Mike Stuhff, interview, September 2004.

Dan Benyshek, interview, September 2004.

David Zimmerman, personal letter, November 2002.

A. Lynn Allison, "The Navajo Witch Purge of 1878," *Arizona State University West Literary Magazine*, May 2001, www.west.asu.edu//paloverde/Paloverde2001/Witch.htm.

Doug Hickman, "Navajo Skinwalkers," www.sohh.com/forums/showpost.php?p=2818513&post count=58.

Tony Hillerman, *Skinwalkers* (New York: Harper, 1986), p. 250.

"Skinwalkers," About Paranormal Phenomena, http://paranormal.about.com/library/weekly/aa061801a.htm.

CHAPTER 7. *CHUPAS*

Simon Harvey-Wilson, "Phasers & UFO Light Beams," http://homepage.powerup.com.au/~tkbnetw/Simon_Harvey-Wilson_6.htm

Jacques F. Vallee, *Confrontations* (New York: Ballantine Books, 1991).

Bob Pratt, *UFO Danger Zone: Terror and Death in Brazil* (Madison, WI: Horus House, 1996).

John F. Schuessler, *The Cash-Landrum UFO Incident* (LaPorte: TX: Geo Graphics, 1998).

National Investigations Committee On Aerial Phenomena, "The Cash Landrum Case," www.nicap.dabsol.co.uk/cashlan.htm.

CHAPTER 10. MUTES

George E. Onet, "Animal Mutilations: What We Know," 1997, www.nidsci.org/ articles/animal1.php.

Frederick W. Smith, *Cattle Mutilations: The Unthinkable Truth* (Cedaredge, CO: Freedland, 1976).

Colm A. Kelleher, *Brain Trust: The Hidden Connection Between Mad Cow Disease and Misdiagnosed Alzheimer's Disease* (New York: Paraview Pocket Books, 2004).

D. Kagan and Ian Summers, *Mute Evidence* (New York: Bantam Books, 1984).

Jacques Vallee, *Passport to Magonia: On UFOs, Folklore, and Parallel Worlds* (New York: McGraw-Hill/Contemporary, 1993).

CHAPTER 13. APPROACH

Erling Strand, "Project Hessdalen," www.hessdalen.org/index_e.shtml.

Massimo Teodorani, "A Long-Term Scientific Survey of the Hessdalen Phenomenon," *Journal of Scientific Exploration* 18, no. 2 (Summer 2004).

Bruce Maccabee, "Strong Magnetic Field Detected Following Sighting of an Unidentified Flying Object," http://brumac.8k.com/MagneticUFO/MagneticUFO.html.

CHAPTER 18. MYSTERY

Richard Feynman, quoted in Peter Sturrock, *The UFO Enigma* (New York: Warner Books, 1999), p. 4.

Jacques Vallee, *Challenge to Science: The UFO Enigma* (New York: Ballantine Books, 1966), pp. 138-39.

The UFO Phenomena (New York: Time-Life Books, 1987), pp. 50-51, 129-30.

Jerome Clark, *The UFO Book: Encyclopedia of the Extraterrestrial* (Detroit: Visible Ink Press, 1998), pp. 653-61.

Zack Van Eyck, "Millionaire Leads Quest for UFO Data," *Deseret News*, October 20, 1996.

"NIDS Investigation on The Flying Triangle Enigma," National Institute for Discovery Science, 2004, www.nidsci/org/articles8_25trireport.php.

CHAPTER 20. MONSTERS

Gayle Highpine, "Attitudes Toward Bigfoot in Many North American Cultures," *The Track Record,* 1992, www.n2.net/prey/bigfoot/articles/highpine.htm.

Ron Murdock, "Sasquatch," www.paranormal.com/ghoststudies/sasquatch.html.

Michael A. Cremo and Richard L. Thompson, *Forbidden Archeology* (San Diego: Bhaktivedanta Institute, 1996), p. 595.

Joyce Bynum, "Bigfoot—A Contemporary Belief Legend," *Et Cetera* 49, no. 1 (September 1992): 352, www.rfthomas.clara.net/papers/bynum.html.

John Keel, *The Complete Guide to Mysterious Beings* (New York: Tor Books, 2002), p. 81.

Wesley Williams, "Bigfoot-Sasquatch FAQ," http://home.nycap.rr.com/wwilliams/BigfootFAQ.html.

Helmut Loofs-Wissowa, "Seeing Is Believing, or Is It?" *ANU Reporter* 27, no. 12 (July 17, 1996), www.rfthomas.clara.net/papers/seeing.html.

Nick Redfern, "Major Conference on Bigfoot in October," *Phenomena* online, September 8, 2004, www.phenomenamagazine.com.

Stephen Wagner, "Are We Closing in on Bigfoot?" About Paranormal Phenomena, www.paranormal.about.com/library/weekly/aa040901a.htm.

Nik Petsev, "In the Footsteps of the Komodo Dragon," www.cryptozoology.com/articles/footsteps.php.

Loren Coleman and Jerome Clark, *Cryptozoology A to Z* (New York: Simon & Schuster, 1999), p. 17.

Tony Healy, "High Strangeness in Yowie Reports," Myths and

Monsters 2001 Conference,
www.herper.com/miscpdf/MMPapers_Aust.pdf.

CHAPTER 22. OTHER HOT SPOTS

Jerome Clark, *The UFO Book*, pp. 58-60.

David Akers, "Report on the Investigation of the Nocturnal Light
Phenomena at Toppenish, Washington," December 5, 1995,
www.cufon.org/cufon/yak1.htm.

Spar Giedeman, "The Yakima UFO Enigma," Psi Applications,
www.psiapplications.com/spar11.html.

Greg Long, *Examining the Earthlight Theory: The Yakima UFO
Microcosm* (Chicago: J. Allen Hynek Center for UFO Studies,
1990).

Greg Long, "The Paranormal at Yakima," *MUFON Journal*, July
1995, pp. 3-12.

Timothy Good, *Alien Contact* (New York: William Morrow, 1991),
pp. 62-71.

Emmanuel Dehlinger, "UFOs: The Military Unmasked," 2003,
www.ovnis.atfreeweb.com/ufos_the_military_unm.htm.

CHAPTER 24. THE MEDIA

Zack Van Eyck, "Millionaire Leads Quest for UFO Data."

Jim Wilson, "The New 'Area-51,' " *Popular Mechanics*, June 1997.

George Knapp, "Path of the Skinwalker, Part 1" and "Path of the Skinwalker, Part 2," *Las Vegas Mercury*, November 21 and 28, 2002.

CHAPTER 25. HYPOTHESES

Etzel Cardena, Steven Jay Lynn, and Stanley Krippner, *Varieties
of Anomalous Experiences* (Washington, DC: American Psychological Association, 2000).

Andreas Reif and Bruno Pfuhlmann, "Folie a Deux Versus Genetically Driven Delusional Disorder: Case Reports and Nosological Considerations," *Comprehensive Psychiatry* 45, no. 2 (March-April 2004): 155-60.

A. V. Ravindran, L. N. Yatham, and A. Munro, "Paraphrenia Redefined," *Canadian Journal of Psychiatry* 44, no. 2 (March 1999): 133-37.

Michael A. Persinger, "The Most Frequent Criticisms and Questions Concerning the Tectonic Strain Hypothesis," Behavioural Neuroscience, www.laurentian.ca/neurosci_research/tectonic_theory.htm.

Michael A. Persinger, "Geophysical Variables and Behavior: III. Prediction of UFO Reports by Geomagnetic and Seismic Activity," *Perceptual and Motor Skills* 53 (1981): 115-22.

Michael A. Persinger and John S. Derr, "Geophysical Variables and Behavior: XXIII. Relations Between UFO Reports Within the Uinta Basin and Local Seismicity," *Perceptual and Motor Skills* 60 (1985): 143-52.

Roy T. Dutton, "A Testable Astronautical Theory for UFO Events," March 2003, www.daviddarling.info/encyclopedia/U/UFOs_Dutton1.html.

CHAPTER 26. THE MILITARY

Junior Hicks, interview, September 2002.

Ingo Swann, *Penetration: The Question of Extraterrestrial and Human Telepathy* (New York: Ingo Swann Books, 1998).

Angela Thompson, "Confidential Report: Ranch Activities," unpublished assessment prepared by Dr. Angela Thompson and the Nevada Remote Viewing Group," January 2003.

Jim Schnabel, *Remote Viewers: The Secret History of America's Psychic Spies* (New York: Dell, 1997).

CHAPTER 27. THE NATIVE AMERICAN CONNECTION

"Brief History About Alamosa & The Valley," Alamosa Visitor Information Center, www.alamosa.org/index.cfm?fuseaction= standard&category Id=6&subCategoryId=1.

Sean Casteel, "Creatures, Human and Otherwise, of Colorado's San Luis Valley," *UFO Magazine*, December 2001-January 2002, pp. 64-67.

"Los Caminos Antiguos," America's Byways, www.rmpbs.org/byways/lca_ancient.html.

Harrison Lapahie Jr., "Navajo Sacred Mountains," www.lapahie.com/Sacred_Mts.cfm.

Tom Dongo and Linda Bradshaw, *Merging Dimensions: The Opening Portals of Sedona* (Sedona, AZ: Hummingbird Publishing, 1995).

Louis L'Amour, *The Haunted Mesa* (New York: Bantam Books, 1987),

Clyde Kluckhohn and Dorothea Leighton, *The Navajo* (Garden City, NY: Anchor Books, 1962), p. 180.

"Chronology of Ute History," Southern Ute Indian Tribe, www.southern-ute.nsn.us/history/chronology.html.

CHAPTER 28. OTHER DIMENSIONS

James Deardorff, Bernard Haisch, Bruce Maccabee, and Harold E. Puthoff, "Inflation Theory Implications for Extraterrestrial Visitation," *Journal of the British Interplanetary Society* 58 (2005): 43-50.

Jacques Vallee and Eric Davis, "Incommensurability, Orthodoxy and the Physics of High Strangeness: A 6-layer Model for Anomalous Phenomena," National Institute for Discovery Science, www.nidsci.org/pdf/vallee_davis.pdf.

Harold E. Puthoff, "SETI, the Velocity of Light Limitation and the Alcubierre Warp Drive: An Integrating Overview," *Physics Essays* 9 (1996): 156.

Harold E. Puthoff, Scott R. Little, and Michael Ibison, "Engineering the Zero Point Field and Polarizable Vacuum for Interstellar Flight," *Journal of the British Interplanetary Society* 55 (2002): 137.

Matt Visser, S. Kar, and N. Dadhich, "Traversable Wormholes With Arbitrarily Small Energy Conditions Violations," *Physics Review Letters* 90 (2003): 2011.

Eric Davis, "Wormholes—Stargates: Tunneling Through the Cosmic Neighborhood," *MUFON International Symposium Proceedings,* 2001.

Jacques F. Vallee, *Dimensions: A Casebook of Alien Contact* (New York: Ballantine Books, 1989).

Patrick Harpur, *Daimonic Reality: Understanding Otherworld Encounters* (New York: Penguin, 1994).

Michael Grosso, *Experiencing the Next World Now* (New York: Paraview Pocket Books, 2004).

John A. Keel, *Operation Trojan Horse* (Liliburn, GA: Illuminet Press, 1996).

John Mack, *Passport to the Cosmos* (New York: Crown, 1999).

Jacques F. Vallee, *Passport to Magonia.*

CHAPTER 29. OUTER WORLDS

Michio Kaku, "Parallel Universes, the Matrix, and the Superintelligence," Kurzweil AI.net, June 26, 2003, www.kurzweilai.net/meme/frame.html?main=/articles/art0585.html.

Colin Wilson, *Alien Dawn* (New York: Fromm, 1998), p. 155.

Beatriz Gato-Rivera, "Brane Worlds, the Subanthropic Principle, and the Undetectability Conjecture," August 19, 2003, http://arxiv.org/pdf/physics/0308078.

Charles Choi, "Universe May Exist in a White Hole," *Colorado Daily,* May 10, 2003.

Mark Baard, "UFO Seekers Search for Respect," Wired News, November 15, 2002,
www.wired.com/news/culture/0.1248,56334-2,00.html.

Max Tegmark, "Parallel Universes," *Scientific American*, May 2003, pp. 44-46.

CHAPTER 30. INNER WORLDS

John Mack, *Passport to the Cosmos.*

Olga Kharitidi, *Entering the Circle: Ancient Secrets of Siberian Wisdom Discovered by a Russian Psychiatrist* (San Francisco: Harper, 1997).

Kenneth Ring, *Heading Towards Omega: In Search of the Meaning of the Near-Death Experience* (New York: William Morrow, 1984).

Richard Strassman, *DMT: The Spirit Molecule* (Rochester, VT: Park Street Press, 2001).

Michael Harner, *The Way of the Shaman* (San Francisco: Harper, 1990).

CHAPTER 31. REVOLUTIONARY SCIENCE

Thomas S. Kuhn, *The Structure of Scientific Revolutions*, 3rd ed. (Chicago: University of Chicago Press, 1996).

Jacques Vallee, *Messengers of Deception: UFO Contacts and Cults* (Berkeley, CA: And/Or Press, 1979).

John Mack, *Passport to the Cosmos.*

"Interview with John Mack," *Nova*, 1996,
www.pbs.org/wgbh/nova/aliens/johnmack.html.

John Mack, "Messengers From the Unseen," *Oberlin Alumni Magazine*, Fall 2002,
www.centerchange.org/passport/oberlin02.html.

EPILOGUE

H. E. Puthoff, "Searching for the Universal Matrix in Metaphysics," *Research News and Opportunities in Science and Theology* 2, no. 8 (April 2002), www.earthtech.org/publications/RNOST_v2_p22.pdf.

Stephen Hawking, *The Universe in a Nutshell* (New York: Bantam Books, 2001), pp. 4, 26.

Colin Wilson, *Alien Dawn*, p. 45.

Marcus Chown, *The Universe Next Door* (New York: Oxford University Press, 2002), p. 97.

Charles H. Lineweaver and Daniel Grether, "What Fraction of Sun-like Stars Have Planets?" *Astrophysical Journal* 598 (December 1, 2003): 2.

Beatriz Gato-Rivera, "Brane Worlds."

Hazel Muir, "Life on Venus?" *New Scientist,* May 4, 2002.

Leonard David, "Scientists Seek Scent of Life in Methane on Mars," Space.com, August 24, 2004, www.space.com/science astronomy/mars_methane_040824.html.

Jim Wilson, "Science Does the Impossible," *Popular Mechanics,* February 2003, p. 63, www.popularmechanics.com/science/research/1282281.html.

David Deutsch, *The Fabric of Reality* (New York: Penguin, 1998).

Max Tegmark, "Parallel Universes," pp. 41-46.

Acknowledgments

We wish to thank the many people who were associated with this project. They made invaluable contributions, of which some were fundamental, some were creative, and all were important. We apologize if we have missed anybody:

Joseph (Junior) Hicks, Chad Deetken, Ryan Layton, Chris O'Brien, Scott Little, Pete Pickup, Joan Pickup, John and Dorothea Garcia, Zack Van Eyck, Dr. George Onet, Mary Allman, Dr. Eric Davis, Roger Pinson, John Velier, Tommy Blann, Gabriel Valdez, George Arnett, Richard and Jean Dietz, Michael Stuhff, Dr. Dan Benyshek, John Peterson, Dr. Angela Thompson, John Schuessler, Dr. James Whinnery, Dr. Douglas Ferraro, Dr. Edgar Mitchell, Dr. Ted Rockwell, Dr. Warren Burrgren, Dr. Johndale Solem, Phylis Budinger, Richard Wilson, Dr. Horace Drew, Dr. Martin Piltch, Dr. Albert Harrison, Dr. Melvin Morse, Dr. Jessica Utts, Dr. Kit Green, Melinda Floyd, Connie Van

Horne, Shane O'Brien, Anne Fechter, Dr. Bruce Cornet, Matthew Adams, Eric Sorenson, Robert Stoldal, Geoff Schumacher, Dr. Jacques Vallee, Dr. Hal Puthoff, Dr. John Alexander, Keith Wolverton, Robert Bigelow, and of course, our ever-patient editor Patrick Huyghe.

Index